Build-It Book
of Miniature Test &
Measurement Instruments

No. 792
$7.95

Build-It Book
of Miniature Test &
Measurement Instruments
By Robert P. Haviland

TAB BOOKS
Blue Ridge Summit, Pa. 17214

FIRST EDITION

FIRST PRINTING—JANUARY 1976

Copyright © 1976 by TAB BOOKS

Printed in the United States

of America

Hardbound Edition: International Standard Book No. 0-8306-6792-X

Paperbound Edition: International Standard Book No. 0-8306-5792-4

Library of Congress Card Number: 75-41720

Cover photo courtesy of Electronic Technical/ Dealer.

Foreword

Basically, this is a how-to book. It describes the design, construction, and calibration of miniaturized measuring equipment useful to electronic experimenters. The instruments covered will be useful to experimenters generally, to amateurs, hi-fi buffs, CB operators, and to computer experimenters and servicemen.

This is also a training text. An experimenter who starts with the simple instruments described first, then proceeds with the later instruments, will develop both construction skill and measurement expertise. If a little time is spent on the principles of operation of each instrument and is followed by some "hands-on" use in practical test and measurement applications, the experimenter will be well on the way to becoming a qualified instrument man.

To obtain these two objectives, the material is organized by chapters, one for each instrument. Each of these chapters covers the working principles of the instrument and its construction, calibration, and use. In addition, chapters on the special techniques of instrument calibration and on construction are included to supplement the data of the construction chapters. However, each instrument can be completed without reference to other chapters.

Adherence to a particular sequence of construction is not necessary; but if you lack instrumentation experience, you'd be well advised to follow the sequence as presented. The instruments themselves increase in complexity from one chapter to the next. The initial instruments, while capable of developing full accuracy, require only simple calibration techniques. In later instruments greater precision is required if full calibration accuracy is to be developed.

Robert P. Haviland

Contents

Chapter 1
Why the Miniature Lab?

Why should you bother to build a laboratory instrument rather than buy the instrument commercially? After all, the commercial instrument has been fully developed, its characteristics are known, and it is available. Then, too, building and calibrating laboratory instruments is a job for a specialist, one who knows what he is undertaking and who can accomplish the work accurately and rapidly. In view of these factors, why bother to build?

There are at least four basic reasons for home building:

- Equipment can have needed features
- Unneeded features can be omitted
- It is good experience
- It is less expensive

Quite often an experimenter or serviceman will have a particular problem repeated many times. When this occurs he can save considerable time and obtain much better results if he takes time to set up a special piece of measuring or test equipment rather than using an assembly of several standard equipments to do the same job. Then too, sometimes he will need a special piece of equipment which is not available commercially. For example, suppose you have an often repeated need for measuring the gain of transistors at a particular frequency. You can easily build up a small combination oscillator, voltmeter, and power supply to measure the transistor gain at *precisely* the desired frequency. While such measuring devices have been described in transistor manufacturers' application notes, experimenters' magazines, and other literature, they do not seem to be available commercially. So you build. Some of the instruments in this book were constructed for the simple reason that they are useful but no commercial counterparts exist.

In addition to obtaining desired features, instruments can be simplified by omitting undesired or unneeded features. For example, if a small radio frequency (RF) oscillator is to be used only for ham-band antenna standing-wave-ratio measurements, the features of calibrated attenuators, modulation, and even continuous tuning range are not necessary. Considerable simplification is possible by constructing a special unit just for the desired frequencies—one with no provisions for output level control or for modulation.

Building your own instruments is excellent experience. It is one of the best ways that you, as an amateur or experimenter, can extend your understanding of circuits and circuit performance. In the process you can develop into a qualified, well trained instrument man. Just for example, if you undertake to build, calibrate, and put to actual use the various instruments and accessories described in this volume, you would be justified in claiming to have experience in precision measurements. And you should have no difficulty in backing up your claim in *any* laboratory, after a short period of becoming acquainted with the particular instruments in that laboratory. If, in addition, you extend the concepts shown here, and design, build, and calibrate some instruments on your own, you should have no hesitation in claiming that you have *design* experience; the instruments you fabricate will speak for you.

The final reason for building your own instruments is economy. In direct cash outlay, you can build a piece of equipment for one-fourth to one-tenth the cost of a new commercial instrument. A further cost advantage occurs if the instrument to be constructed is a special-purpose type. Often, the investment can be minimized by judicious use of surplus and junkbox components—at times even to the point where the instrument represents virtually no cash outlay. And yet, by taking pains in construction and calibration, results can be made equal to that which would be obtained with commercial equipment.

THE BUILD-OR-BUY DECISION

How does one tell whether to build from scratch, buy a kit, or buy a ready-made instrument? It is pretty much a matter of looking at the relative merits favoring each choice, deciding whether the balance is in favor of *build* or *buy*. For example, if suitable commercial test equipment is available, it would not be wise for a busy service shop to take time to design and build a piece of equipment—even taking time to assemble a kit would be questionable. On the other hand, a beginning experimenter working on a tight budget could find a piece of commercial equipment (kit *or* complete) a burdensome investment. If the experience of building is a factor to consider, by all means think of it as a weighty element. If you already have all the experience you need, then you should take the "time is money" attitude in your decision-making process.

WHY MINIATURE?

I have built measuring equipment of various types for many years, without paying particular attention to size or to appearance. Most of these instruments were built to obtain a needed feature or to save money for some other investment. This situation changed when I decided to undertake amateur radio operation aboard a boat in a serious way. Boats don't have much space, nor did the particular sailboat selected have much in the way of power. On the other hand, doing without measuring equipment was not an acceptable alternative. This led me to the design of several small, battery-operated instruments. These functioned so well and were so useful that I decided to extend these to a reasonably complete laboratory. In the process, I devoted time and thought to appearance and uniformity, which led to the adoption of a design format that could serve as a thematic element of style which could be applied again and again.

A word about the instruments themselves. While these are intended for use by the experimenter in his laboratory work, the instruments cannot be called *laboratory grade*. General Radio and Hewlett-Packard are not going to be put out of business—nor are the kit manufacturers, even though the instruments described are essentially of the same quality reflected in the pages of Heathkit and Eico catalogs. The instruments described here are slanted to the experimenter, and especially the experimenter on a tight budget.

For a design to be included in this book, several factors were considered. One was that experience had shown that a need exists. Further, the instrument type had to be amenable to home construction and to miniaturization. Then too, a goal was to keep the cost to a fraction of that of commercially available instruments of equal quality. By judicious use of surplus components, it should be possible to build any of these instrument circuits for less than $30.

All of the instruments described here have been built, tested, and used extensively. All have gone through the process of breadboard, first prototype, and clean prototype. The results, as far as appearance goes, can be judged from the photos in this book.

Chapter 2
Avoiding that Home-Built Look

A problem which plagues home constructionists of electronic equipment is lack of a "professional" appearance in the finished job. Too often the results appear as in Fig. 2-1—extra panel holes, poor layout, poor finish, paste-on lettering, and so on. The contrast with commercial or kit equipment is just too much.

This appearance problem is almost totally the result of only one factor—attempting to save time. The at-home builder is often in a hurry to get the project finished, or to go on to another. In the case of special test equipment, the usual goal is to get back to the basic project for which the equipment is intended. Yet a home builder, using very simple equipment, can do a very creditable job of maintaining appearance. Figure 2-2 shows examples of three instruments, some of the series described in this book. Attaining good appearance was a definite goal. These instruments involve an expenditure of about a 10% increase in construction time and a few cents of extra cost per instrument. The additional construction time is that required to convert from a prototype to a dress sample: time spent in layout, cleaning, painting, and especially in making direct calibrations. The additional cost is due to a few squirts of paint from a spray can and a few letter transfers from a panel-lettering kit. These are mighty small investments and seem well worthwhile.

Let us look at the steps required to turn out a good appearing piece of equipment.

THE COMMERCIAL DESIGNER'S SECRET

The real secret of commercial equipment appearance is that the designer has taken several steps beyond those usually

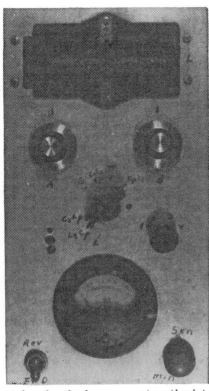

Fig. 2-1. Typical results of a home project. This antenna tuner shows the defects of extra holes, mismatched dials, freehand lettering, and poor panel finish.

taken by the home constructionist. His early steps are just the same: first, the breadboard, always messy, made with tacked-on components barely hanging together. It looks precisely like an experimenter's rat nest, the one difference being that it is usually surrounded with high-priced measuring equipment. The commercial designer follows this with a prototype, packaged in a cabinet of appropriate size, with controls and leads in about their final position, and with all functions labeled. This does not look much different from a similar unit prepared by a home experimenter.

This is usually the point at which the home builder stops—and the point at which the commercial designer secures additional help. The commercial equipment will go through at least one additional step, often involving work by specialists in mechanical construction, control placement, and appearance (*human* engineering, it's called). The end result of this additional step is a production prototype. This is a hand-built unit, almost indistinguishable from production units. It is sometimes called a dress sample, by analogy to a dress rehearsal in the theater.

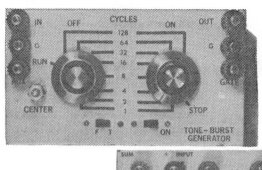

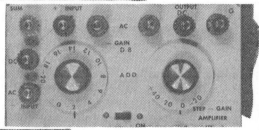

Fig. 2-2. Appearance attainable with a little extra time and effort. The two larger units are finished with black pressure-transfer lettering on a white panel finish. The calibrated dials of the center unit are homemade.

The amateur or experimenter usually does not need to go to the point of scrapping the prototype and building a dress sample from scratch. With a little care in planning and construction, his rough prototype can be converted to a dress unit, at least as far as external appearance is concerned. Even the internal appearance can be kept neat and eye-pleasing with just a little care.

The major elements in securing a good external appearance are the layout, the general finish of the instrument, the panel lettering, and details such as dials and dial calibration. Internally, layout and construction, neatness of cabling, and arrangements for supplying power are major factors. Let us look at each of these.

OVERALL FINISH

The simplest method of securing a uniform overall finish is to use bare aluminum. There are several forms which can

15

be used. The first is "as supplied" metal, which is not especially attractive; it's easily scratched and shows finger marks readily. Figure 2-1 shows this type of finish.

A variation on the bare-aluminum theme, based on the fact that aluminum is easy to scratch, is the *brushed* finish. In this, the surface, usually only the front panel surface, is scratched with a fine-bristled wire brush, moved with long parallel strokes. Usually it is best to have the brush marks go in the long direction of the panel. This brushing should be continued until a uniform surface has been developed: the surface should appear the same for light striking from any direction.

A brushed-surface instrument face is shown in Fig. 2-3. Note, however, that some of the letters are scratched, and the ON marking is nearly obliterated. This points up the need for the application of a good, tough transparent lacquer—particularly over lettering that is likely to be rubbed by the fingers in actual instrument use.

A third bare-aluminum treatment, usually called a satin finish, is obtained by etching the panel with lye. For this, the panel must be very clean and free of grease—especially finger marks. If this precaution is not taken, the panel will not etch uniformly and will have an appearance that is anything but pleasing.

All of these bare-metal finishes can be improved by coating with a transparent preservative. This prevents deterioration and makes it easy to remove dust and finger marks. Acrylic enamel is a good preservative coating.

After many trials, I've settled on painted panels and bare-metal cases as my standard finish. This gives a uniformity in appearance which is pleasing when several

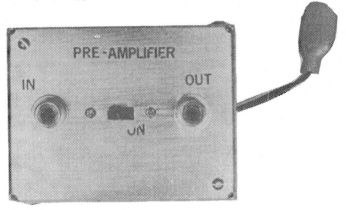

Fig. 2-3. Brushed aluminum finish, made with a wire brush. After lettering, the finish is protected with a coat of transparent lacquer.

instruments are used together. Paint color is a matter of personal preference; flat white seems well suited for small instruments, as shown by two of the examples of Fig. 2-2. Pastel colors seem appropriate for medium-sized packages, and darker colors for those with large panels. The range of colors available is almost unlimited. There are also paints controlled to a certain shade, such as the tradename paints (*Chrysler* blue, *Caterpillar* yellow, *Coca-Cola* red, etc.).

Finishing the panel with paint is relatively simple, even if the unit has been previously used for prototype development. The preferred process is to remove all components from the panel and then clean it with fine steel wool (the type that has soap inside the wool). After making sure that the panel is dry, a base coat is then sprayed on. With flat white it is relatively easy to secure good coverage with a thin coat which does not tend to chip. When the panel dries, apply the transfer lettering below and cover with a transparent coat of a compatible paint. The acrylics seem to work very well for both the base coat and the transparent coat. With care, colored lettering or two-colored panels can be secured.

Nearly as good as paint for temporary use is the use of self-adhesive decorative plastic films. This is prepared by first applying a base layer of a light color film to the panel. Start with a piece of plastic which is oversized and trim it to the desired size after it has been placed on the panel. At this time, all cutouts should be cut through the plastic, using a sharp knife or a razor blade. As sold, the plastic is covered with wax. This should be scrubbed off, using a gum-type eraser. This will change the shiny appearance of the plastic to a dull appearance. When cleaned, lettering should be applied to the panel. When this is done, a piece of transparent plastic should be placed over the panel, again one cut oversize. This should be trimmed to just cover the first layer. The general appearance of this method of finish is quite good and highly durable. There is, however, a tendency for the plastic to pucker around screw heads and shaft mounting nuts. This can be avoided by cutting a hole slightly larger than the screw head or mounting nut, using circular punches available from most hardware stores.

PANEL LETTERING

The next most important factor in securing good appearance is the lettering itself. This is far, far easier than it used to be, thanks to the pressure-transfer, self-adhesive lettering sold under such various tradenames such as *Datak, MarKit, Instant Lettering*, etc. (available in electronics distributors, art stores, and office-supply outlets). Commonly available colors are black and white; sometimes gold and silver can be found, but these seem to be more readily available in the form of decal transfers than pressure transfers. The various types can be mixed if desired. Bear in mind that *none of these lettering systems is durable unless protected by a coating of transparent lacquer.*

There are some precautions needed in using these lettering sets. First, the sets with made-up words are easier to use than the sets with individual letters, since automatic alignment and spacing is secured. They seem to be a good buy, even though many words will never be used. A fresh letter kit (they tend to dry out from age) is essential for use on metal. Because of this, it is desirable to buy small kits rather than the large ones. It is also a good idea to keep the letter kit in a plastic container and in the refrigerator.

In applying the letters, watch the alignment closely. Make sure the letter is in the right place before applying the burnishing tool. Watch also for interference with binding posts, knobs, etc. Obviously, appropriate-sized letters should be used: large for equipment titles and for the commonly used controls, small for other purposes. When a word has been completed, it is a good idea to place one of the separator sheets (they come with the letter sets) over the area and to burnish the word a second time. This makes for better adhesion and lessens the chance that a letter will be picked up as the rest of the panel is labeled.

Correcting an error is simple—remove the bad lettering with a fingernail or toothpick, wipe the area clean, and start over. When the lettering has all been finished, wipe the entire panel very carefully with lighter fluid or other cleaning fluid which does not affect the lettering or plastic stick-on film, if used, then spray the panel with transparent lacquer. It should be good for years of service.

LAYOUT

By and large, symmetrical layout of equipment panels is best. However, don't follow this slavishly by forcing internal layouts into poor designs. Nonsymmetrical layouts do not look that bad, and, in fact, can be quite pleasing. An artist will automatically place the various components in attractive locations. If you have no flair for art, use the principles of dynamic symmetry as a guide to layout; a number of art books describe these principles.

It should be possible to obtain pleasing appearance and good operating and construction characteristics simply by placing the panel parts on a full-scale panel drawing. It is usually desirable to have inputs to the left, outputs to the right. Meters are generally best at the top of the panel area and controls in the center and the bottom area. Adequate space should be left around knobs and terminals. For binding posts, allow ½ to ¾ inch. Jacks can be closer spaced. Finger space clearance is needed around knobs; however, many indicating dials and the indicator part of knobs can be spaced much closer—almost in contact.

DIALS

One of the areas where commercial equipment has a marked edge over home-built equipment is that of dials and dial calibration. As long as simple and arbitrary scale markings are all that is needed, you can find a variety of good-looking, low-cost dials. But if you need a special marking or a nonlinear scale, the commercially available knobs are of no help—they just are not available in special markings. True, there are a few commercially available dials intended for direct calibration of receiver tuning. But without exception, all of these are much too large to be used on miniature instruments. In fact, most of them require that the layout of the entire receiver start with the dial.

In an attempt to rectify this situation, I spent considerable time on the development of dials, indicating markings, and methods of direct calibration. The following describes several of these methods—the ones which I found most practical.

Let us look first at variations of simple knobs and dials, such as those in Figs. 2-4 and 2-5. A common knob is the black pointer found on much of the older equipment. For its intended purpose, it is perfectly satisfactory, but it does look somewhat dated. A modern version can be home-built by taking a small aluminum knob, obtainable from the retail supply houses, and cementing a small aluminum pointer to this with epoxy. A variation is to scribe a mark across the top of the knob in alignment with one on the side.

Also shown in the figures are skirted knobs, in both commercial and home-built versions. The commercial version with the indicator bar is really intended for volume and tone control use. However, it works very well with numbers lettered on the panel. These may be uniformly spaced, as for switch positions, or nonuniformly spaced, if required by the calibration. A variation in this design has a numbered calibration around the skirt. While this knob is neat enough, it is rare that desired markings coincide with those on the knob.

The home-built knobs in Fig. 2-5 and 2-6 are usable for either uniform or nonlinear calibration. The one with the

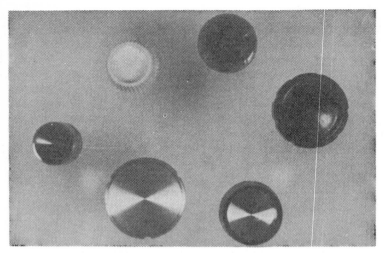

Fig. 2-4. Small knobs suitable for miniature equipment. The white unit is constructed from a toothpaste cap, with a bushing of one-eighth inch (ID) copper tubing cemented in, then drilled and tapped for a 6−32 setscrew. It is for ⅛ inch control shafts.

transparent skirt is intended to be placed above markings on the panel. It is made by attaching a transparent plastic disc, obtainable from hobby supply stores, to the back of a plain (but small) knob. Fastening may be by epoxy glue, but greater durability will be obtained if the plastic disc is held in place with three 2−56 screws tapped into the plastic knob. The

Fig. 2-5. An assortment of pointer-type knobs. The unit with the transparent skirt and opaque line is homemade, as described in the text.

fiducial mark is a piece of drafting tape made for printed circuit work. It is placed on the back of the transparent disc.

An inverse treatment of this transparent skirt knob design (not shown) is to cover the skirt with an opaque paint or plastic film, leaving a window at the fiducial position.

The knobs of Fig. 2-6 use the same basic construction. However, before the transparent disc is assembled to the center knob, it is covered with a film of opaque plastic of appropriate color, or painted (white is a good choice). The desired markings (index marks and numbering) are then placed on this plastic or paint film. After marking is complete, the surface is protected by transparent paint or film. The knob is then assembled, preferably using screws to hold the skirt to the knob. Remember to scrub the first plastic film with a soft gum eraser to remove the wax from the plastic before attempting to transfer the lettering.

Another useful technique is to place a small plate under the mounting nuts holding the particular control, as in Fig. 5-1. The plate carries the calibrations. A number of such plates are available commercially for switch calibrations and relative volume indications. You can readily fabricate your own as an alternative to placing calibration marks directly on the panel. This has the advantage of permitting easy recalibration should it ever become necessary.

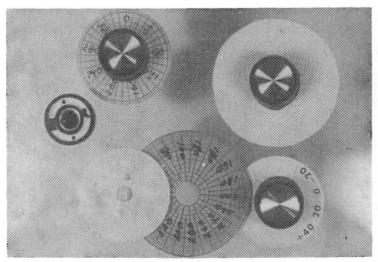

Fig. 2-6. Materials for preparing custom calibrated dials. The two discs with polar coordinate graph paper faces are used to determine the calibration points. The knob shows the method of filling the rear of a standard unit with epoxy to allow attachment of a blank skirt. An assembled, plus an assembled-and-calibrated dial are shown.

Plastic, aluminum, and copper discs suitable for these dials or dial plates can be obtained from hobby stores. The finish can follow exactly the techniques of finishing the panel—bare metal, a plastic film, or paint covering. The painted metal plates are the most durable.

The calibration procedure for these knobs and dial indications is as follows. First, make up a dummy dial or dial plate with a piece of polar graph paper to allow angles to be measured. Such dials and plates are shown in Fig. 2-6. Proceed with the calibration process as described in the next chapter, recording the dial position in degrees from a reference point. When the calibration values have been secured and checked, lay the blank dial face or dial skirt over another piece of graph paper. It can be held in place by tape or rubber cement. Mark the outer edge of this plate at the appropriate angles. A ballpoint pen works well for the plastic film. (Mark lightly; after it dries, scrub it with a soft eraser.) A pencil works well on paint, or India ink can be used. Markings can also be made from short pieces of printed-circuit drafting tape. Numbers are placed on the disc as needed. If desired, the knob or skirt can be labeled with the name of the variable it controls. The lettering and markings should be protected by a plastic film or transparent lacquer. When this work is completed, the dial is assembled and installed in place of the temporary unit.

INTERNAL APPEARANCE

There is no doubt that the development of printed circuit (PC) boards has simplified the problem of the home constructor in obtaining good internal appearance of his equipment. Virtually all of the projects described here use these boards. All of them use single-sided boards, which does present a small problem in lead layout when many integrated circuits are used, since it is almost impossible to prevent lead crossing. For this reason a number of the projects are almost a combination of printed circuit boards and point-to-point wiring techniques.

All of the boards described here have been prepared using hand layout methods. Many of the board layout drawings are also made by hand; others have been prepared with commercial drafting aids.

The hand-layout technique I use is to first place the parts in approximately their correct position. A few pencil sketches may be made at this time. This is followed by a layout on graph paper, usually in two colors—one showing the component positions, the other the interconnects. When this layout is completed and checked, I fasten it to the foil side of a suitable piece of circuit board. Then I mark the drill points with a pick or punch. (The layout graph paper is then removed and used as a guide in masking the board.)

I draw in all soldering pads and leads with an ink-resist pen. Large nonetched areas are outlined rather than completely filled in. I inspect the board for layout correctness and for leads which are too close together. After corrections are made as needed, I fill in the large nonetched areas using fingernail polish. A layer of fingernail polish may also be applied to the larger conductor sections and to the soldering pads as a precaution against erosion of the paint resist. The board is now allowed to dry, etched, and the resist removed using lacquer remover. Drill the board using first a No. 60 drill for all holes, then appropriate larger drills for mounting points, controls, etc.

Of the various types of board available, it seems best to standardize on glass—epoxy single-sided board. This is typically available from surplus sources at an attractive cost. The difficulty with paper-filled plastic boards is warpage. Some other types are excessively brittle. The glass—epoxy board has very few defects, but it does tend to wear out drills rapidly.

POWER SUPPLY

All of the instruments described here have been developed to work on battery power. Most of the instruments work directly from standard 9-volt batteries. The ones using operational amplifiers use two such batteries. Many instruments will work equally well on a 12-volt supply—some directly, others through a regulator circuit. The designs which use digital integrated circuits require a +5V source. For battery operation, it is best to obtain this from a 12V source, since the drain is too great for the small 9V batteries.

Although all instruments have been designed for battery operation, all have also been checked using an AC supply. The power supply used is very simple, as shown in Fig. 2-7.

In a few of the instruments, where portability or signal leakage is important, the battery supplies are internal, but most of the instruments are intended for use with external batteries, connected with pigtail leads. Figure 2-3 shows the arrangement. This is not the neatest method, but it is quite practical: it does not require too great an investment in batteries and it permits recharging of the batteries from time to time. The systems using a +5V source should use a different type of connector; the approach I adopted is to use standard TV antenna lead-in connectors, but this idea could probably be improved.

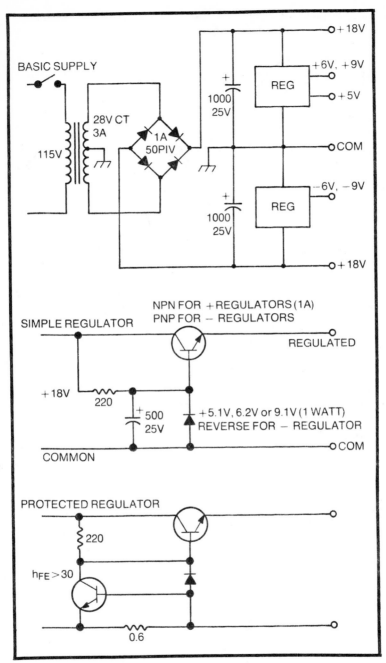

Fig. 2-7. Basic power supply, usable with all instruments of the miniature lab series. Either regulator may be used.

If battery operation using pigtail connections is not desired, mount a miniature plug on one end of the instrument and a mating socket on the other. A standard pin assignment should be adopted, with a different pin for each voltage. Within the instruments all pins should be assigned uniformly. Using this technique, a power cable can be plugged into the first one, and so on. This will give a neat and usable system. Of course, another possibility is to use small individual battery eliminators for each instrument. Most of those on the market are well filtered and can be used directly. If not, a small regulator/filter can be installed within the instrument case.

CORRECTION OF MISTAKES

It is almost certain that you'll be faced eventually with the problem of correcting an error in design or construction, or making a repair. This is more likely if you're designing your own equipment, but it can (and too often does) occur even if a prepared design is being followed. Here are some hints which may prove to be helpful:

The most common problem will probably be a gap in the printed circuit wiring, due to a flaw in the etch resist. This can be corrected by soldering a "jumper" strand of fine wire across the gap. Usually the jumper will have sufficient capillary attraction to pull solder completely across the PC-path gap.

Another common fault is an area of pulled-loose foil—usually a solder pad. Sometimes it is possible to work this back into place using a drop of epoxy to hold it there. If this is not possible, another technique is to cut the defective foil away completely and replace it with an adhesive-backed terminal (sold under the name of *Circuit Stik*). The accompanying instructions show the technique of making connections to the rest of the foil.

A common trouble encountered in design is layout reversal. This is especially likely if a new type of component package is used for the first time. If this happens, it may be possible to mount the component or its mounting socket on the reverse side of the board from that originally intended. If this is to be done, it is a good idea to place solder pads on the insulation side of the board, using a *Circuit Stik* terminal as a bond. This will help in preventing the foil from being pulled loose by pressure on the component.

Another way of handling this reversed layout problem is to make up a corrected layout on a small piece of board and then wire it into place, using jumpers as needed.

ADDING COMPONENTS

In many designs there is the problem of adding components, perhaps to extend the performance or simply to obtain the performance originally desired. Where possible, try to leave space for at least one additional transistor or one additional integrated circuit. These can be added, if needed, by using *Circuit Stik* soldering pads. Almost all layouts shown in this book include additional drill points, and especially ones through critical leads. These are made available to allow addition of external test points, modification to switching, etc. Sometimes a gap in the wiring is intentionally left, with a small jumper being indicated. This allows for a future change to a different type of component, or for interrupting the circuit by a switch or addition of a component. Leave a small pad area in which an adjustable potentiometer can be mounted, where appropriate. This is good practice when integrated circuit (IC) operational amplifiers are used, especially surplus devices. Many times such a trimmer will allow an otherwise useless op-amp to function with perfect satisfaction.

The problem of deletion of components is usually simple. Most of the time it is only necessary to remove the component. Sometimes a small jumper will be needed. Only occasionally will removal of wiring or conductor area from the board be needed; the usual reason is stray coupling to some other circuit.

EXTERNAL CHANGES

Externally visible mistakes can often be repaired also. For example, a misplaced hole can be filled by a dummy screw. Larger holes can be covered with the snap-in plates available. On the front panel, judicious use of dial plates instead of direct marking on the panel can be used to cover relatively large holes. If necessary, a complete false front can be made for the instrument; this has the additional advantage of being easier to work with during lettering and finishing. If the false-front panel is to hide many extra holes, it is best to make it of relatively thick stock, and to cut away the old panel, except for a small mounting lip. If only a few noncritical holes are to be covered, or if the front panel is only scratched, the false front can be a thin piece of material.

You may wish to experiment with panels made from a piece of PC board. This can be etched to give a two-tone finish, or it can be etched to leave dial marks, lettering, etc. The finish can be left in copper, polished, and covered with a protective layer of transparent paint, or the copper can be nickel- or silver-plated.

TOOLS

If you're starting to work with PC techniques or miniaturized equipment for the first time you will find a number of special tools to be most helpful.

A magnifying glass is almost essential. The best, of course, is the type mounted on a movable arm and surrounded by a Circline fluorescent lamp. These are expensive and a substitute may be required.

Another desirable tool is some type of desoldering device. The kind using a small rubber bulb with a Teflon tip is satisfactory, but the type built into the soldering iron is somewhat easier to use.

I find it impossible to work without at least one pair of medical forceps—called *hemostats*. For miniature work, the small five-inch hemostat is best. The forceps sold by radio stores are often too large and clumsy for miniature work, so make an effort to buy medical surplus material. Still another useful nonradio tool is a nail clipper, the type which is an end nipper. This is ideal for clipping component leads after board soldering.

The two most satisfactory ways of cutting PC boards to size are the use of a nibbler and a high-speed jigsaw. The *vibrating* (nonreciprocating) type of jigsaw that makes a loud buzzing noise is entirely satisfactory. (Use a fine-tooth blade.)

A number of drills will be needed for PC boards. Usually, No. 60 is the smallest size which can be obtained without special order. For many component leads a No. 63 drill is more desirable; and if much work is to be done, consider ordering several of these. Board drilling should be done with a drill press under good illumination; hand drilling is possible but tedious. Use of standard electric drills for board work is not recommended, since the drill is too liable to "walk," tearing the foil loose.

In soldering PC boards, at least three hands seem to be needed. A special circuit board holder, or a vise of the type having a vacuum base, is a very worthwhile investment. Good light coming from at least two directions and covering the work area is also a virtual necessity. Other than for these factors, the common tools of the home workshop are all that is needed.

COMPONENT MOUNTING HINTS

Some of the boards in this series are laid out to mount components parallel to the board and in contact with it. In others, where space is critical, the component is mounted at nearly right angles to the board, one lead being longer than the

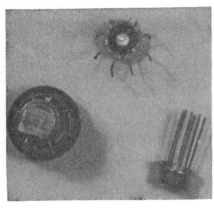

Fig. 2-8. Suggested method of forming leads of TO-5 integrated circuits, by bending over a half-inch-diameter tubing. The large spacing simplifies layout and construction.

other. No special precautions are needed for either arrangement.

In the equipments described, integrated circuits in TO-5 cases are mounted to a half-inch-diameter circle. The method of forming the leads can be seen from Fig. 2-8. These ICs are soldered in. Plug-in mounting is used for dual-inline-package (DIP) units, using sockets or Molex terminals. The sockets are less subject to damage, and less likely to bend the IC leads, but they are more expensive. (DIPs *can* be soldered in place, of course.)

The single-deck switches sold in many radio stores at low cost are intended for point-to-point wiring. They can be converted to PC board use by cutting and forming the terminals as shown in Fig. 2-9. Be sure to make the cut on the same side of each terminal to maintain uniform spacing of the pads.

Fig. 2-9. Method of forming switch contacts to allow use on a PC board. The prongs should all be on the same side of the lugs to allow uniform spacing of mounting holes.

Chapter 3
Measurement and
Measurement Accuracy

Almost everyone measures things, usually many times a day. We measure the speed of a car, the distance to a stop sign, the weight of a letter, or the amount of sugar put into a cup of coffee. But despite the many measurements made, few of us have taken time to understand the process of measurement—and even fewer have developed the understanding and skill needed to make measurements precisely.

One of the main functions of a laboratory is to permit measurements to be made under controlled conditions. Another function is to cause the measurements to be repeatable—to assure that results can be duplicated by another person at another place. Yet another function is to make the measurements possible by providing the measuring tools and processes needed. These functions are just as true for the home experimenter's laboratory as for the giant industrial or government labs.

To accomplish laboratory-grade work, we must go beyond the measurement level of day-to-day events. We need to understand the nature of resolution and accuracy and the importance of such factors as purity and stability. We must develop the skills of calibration and standardization and the ability to use instruments properly. The attitude which leads to use of screwdrivers as pry bars does not go well with instruments.

Those of us who wish to keep expenditures reasonably low must create our own instruments and standards. We must substitute time and skill for money.

In the following pages, the major factors which enter into measurements are reviewed, with special emphasis on the needs of the home experimenter. The discussion is not intended to go to the point of *standard*-precision measurement, which is a way of life in itself. It is intended to be ample for most of our experimentation and serve as an introduction to advanced work.

RESOLUTION AND ACCURACY

To begin a discussion of instruments and measurement accuracy, let us fix some ideas. Suppose that a frequency counter (such as that described in Chapter 11) is used to measure an unknown frequency. The counter is a 5-digit unit and can indicate to a count of 99999. It will measure either in hertz or kilohertz. The nominal error of the counter is plus one count of the least significant digit.

Suppose the counter (I'm actually talking about the one in Chapter 11) is used to read the frequency of an unknown, and that it gives a reading of 04000 on the *kilohertz* scale and of 00234 on the *hertz* scale. What is the unknown frequency? The first reading states that the unknown is between 4000 and 4001 kHz. The second reading says that the last digits are between 234 and 235 Hz. Thus, the indicated frequency is 4 000 234 Hz or, alternatively, 4000.234 kHz.

This is a very impressive measurement figure, but is it the true frequency of the unknown? The answer is, not necessarily! To see why, it is necessary to look first at the method of measurement. The counter works by converting cycles into pulses. It counts the number of pulses which occur in a given period, the *gate* period, where the gate signal is derived from a crystal oscillator. If the crystal oscillator has a frequency of exactly 1 MHz, the gate period is exactly one second or one millisecond long, and the number of pulses indicated by the counter is exactly the number of pulses or cycles per second or millisecond. But suppose that the reference oscillator frequency is in error, say by being 1% low. This will cause the number of pulses (cycles) indicated by the counter to be the number in 1.01 seconds or milliseconds, since the gate is longer than what it should be. The error in the reference oscillator has been transmitted as an error in measurement. Although the digital frequency meter can always tell the difference between two counts, say between 4 000 000 pulses and 4 000 002 pulses, this does not give any guarantee that either reading is correct as cycles occurring during a 1-second period. A digital frequency meter has high resolution, but not necessarily high accuracy.

If we are to secure a correct measurement of the frequency of an unknown, we must have a *transfer standard* of comparison to determine and correct for the error in the count period. In general, all instruments must be related to such a standard for measurement. Suppose we do this for the digital frequency meter by tuning to radio station WWV (U.S. official time standard) at 5 MHz, using a tuned RF amplifier. Attaching the frequency meter just ahead of the detector allows us to measure the carrier frequency of WWV. We can then adjust the frequency of the oscillator in the frequency meter until the indicated frequency reading of WWV is 5 000 000 hertz. We have now standardized the frequency meter, in essence turning it into a *secondary transfer standard*. The measuring period is now precisely the design value.

PURITY AND STABILITY

Suppose we now repeat the measurement with the standardized instrument. Now is the indicated frequency the actual frequency?

Unfortunately, the answer is again, not necessarily. Suppose the signal being measured is frequency-modulated, varying between 4 020 000 Hz and 3 980 000 Hz. If the modulating wave were symmetrical, the average frequency would be 4 000 000 Hz. Since the counter measures the number of cycles which occur during the gate period, it actually measures the average frequency during this period. If the modulating frequency is varying very rapidly with respect to the length of the measuring period, the counter will measure this average frequency and will measure it correctly. However, if the modulating frequency is low compared to the frequency corresponding to the sampling period, the error can be quite large. This is illustrated in Fig. 3-1A, which shows the way the counter will measure an *average* frequency, *which is not necessarily the average of the modulated signal.*

Another effect will show if the modulating frequency is nearly equal to the frequency corresponding to the gate length, or is nearly an integral multiple of this. The counter reading will show a cyclic variation due to the beat between the modulating frequency and the gate frequency. This is called *aliasing* and should be suspected whenever the counter indication varies from sample to sample.

Another reason the measurement may not be trustworthy is shown in Fig. 3-1B. The actual frequency may be drifting, a common phenomenon when equipment is first turned on. It is usually due to temperature change within the equipment. Successive measurements are needed to detect the drift and the stabilized value (as shown by the curve of Fig. 3-1B).

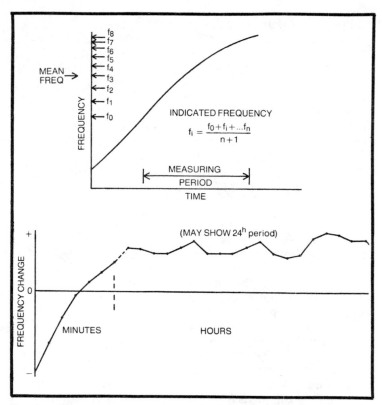

Fig. 3-1. Illustration of the problems of stability and purity in precision measurement—in particular, in the measurement of frequency. (A) If the frequency varies appreciably during the measuring period, frequency will be a mean, approximately related to the actual frequency by the equation. (B) Typical drift curve, with rapid initial change as equipment or measuring instrument warms up, then slower long-term changes, which may be related to such factors as environmental temperature or line voltage.

As a result of these examples we now have some appreciation of four major factors which enter into precision measurements. One of the factors is the *resolution*, or the ability to distinguish between one value and another. The digital frequency meter is inherently an instrument of high resolution. Other instruments, in which the values must be read from a dial or a meter, have lower resolution. Typically, the resolution of the digital frequency meter will be around one part in 10 million, whereas the resolution for a dial or meter will be a few parts in a hundred!

We have also been introduced to the effects of purity and stability of the unknown. Equally, of course, purity and stability are important factors for the measuring instrument.

Finally, we have found the need for an accurate standard of comparison, to allow determination that the reading of the instrument is exact—that the instrument is accurate.

We can see all of these factors at work if we visit a standards laboratory. Suppose we do this, and ask to see a standard ohm. We will be shown a large tubular object with heavy leads, suspended in a bath of oil. The bath is temperature controlled to a fraction of a degree. The leads from the object go to heavy connectors, connection actually being made by a pool of mercury. The object is probably labeled with a statement that it is, in fact, a 1-ohm resistor.

If we ask "How do we know that this is one ohm," we will be informed that it is actually 1.00113 ohms, or some such value. We also will be informed that this value has been checked annually by shipping the unit to the National Bureau of Standards. There it was first given time to stabilize from the strains of shipment, and of being at various temperatures; then it was checked against the *primary* standard of the Bureau. Its measured value was recorded and certified. On return, after allowing it to stabilize again, the secondary standard was cross-checked with other 1-ohm standards, to see if there had been any change during the recalibration. Over a period of perhaps 20 years, this unit has been an ohm. There is every expectation that it is an ohm today and that it will be an ohm tomorrow.

During the visit we can also see the way this precision secondary transfer standard is used. It is never used for measurement of an unknown resistor. It is a reference standard, and its only function is to provide a check of other one ohm units. These other units, the working standards, are the ones which are used to measure unknowns.

STANDARDS IN THE HOME LABORATORY

To apply these factors to the home laboratory we need four things:

- Instruments for making measurements
- Working standards
- Reference standards
- Calibration procedures

A set of instruments is described later, and the actual calibration procedures for each instrument are described with it. For now, let us consider working and reference standards for the home laboratory.

The first thing we should attempt to develop for the home laboratory is a set of reference standards. These do not need to be superaccurate, or even of the quality used in a regular laboratory. However, they should have accuracy consistent with the type of work being attempted. Further, they should only be used as reference standards. They should never be used to make a measurement of an unknown—they are to prepare and check other standards. In particular, it is a good idea to mount the reference standards in a special case and to label them, so that they will not be used accidentally for some purpose other than reference.

All actual measurements of unknowns should be by reference to working standards. These do not need to be separate components. They can be built into an instrument. Of course, it is convenient to have a few separate working standards, perhaps for use from time to time to check instruments or to assemble into a special measuring setup.

There are basically five ways the home laboratory can get a set of standards:

1. Buy as a calibrated standard
2. Buy to marked values
3. Buy or make, as a component, calibrating against standardized units
4. Buy or make, as a component, self-calibrated
5. Buy or make, used uncalibrated (for *relative* comparison)

A good way to start is to either obtain a few components of good accuracy or determine the value of good quality components in a nearby calibration laboratory. This will give the basic standards needed. The remainder of the work can depend on self-calibration, by transfer from these basic standards. Relative calibration should be used only as a last

Table 3-1. Standard Frequency Signals Available.

Frequency	Station
2.500 MHz	WWV
3.333 MHz	CHU
5.000 MHz	WWV
7.335 MHz	CHU
10.000 MHz	WWV
14.670 MHz	CHU
15.000 MHz	WWV
20.000 MHz	WWV
25.000 MHz	WWV

CHU—Ottawa, Canada

WWV—Boulder, Colorado
(also WWVH, Hawaii)

resort. (Still, a relative calibration is better than no calibration at all.)

In principle, only four basic standards are needed. Three of these are from the world of mechanics: *length*, *mass*, and *time*. For electrical work, one additional basic standard is necessary, and can be for any of the electrical quantities. In practice, it is best to have a standard for each commonly used quantity. This means that the home electronic laboratory should be able to calibrate by reference to standards for frequency, time, voltage, resistance, capacitance, and inductance. In addition, some experimenters will wish to have calibration capability for light intensity, acoustic intensity, current, temperature, and for length and mass.

FREQUENCY STANDARDS

Thanks to WWV and CHU, frequency standardization is relatively easy. Instantaneous accuracy of one part in 10^6 is easily obtained, and one part in 10^7 is not difficult. Long-term accuracy of one part in 10^9 or 10^{10} is possible, although very few home-experimenter laboratories have equipment of the stability needed to use this order of accuracy. See Table 3-1 for standard frequencies available.

In most parts of the United States, the power-line frequency, over a long period of time, is essentially as accurate as WWV, since it is corrected to zero average time error. However, at any given instant the power-line frequency may be considerably off, perhaps by one part in 1000 or more. It is very useful as a check reference, however. A further high-accuracy source is the color-burst signal from a TV receiver. The live network-generated color programs use atomic reference standards for this signal, and approach the accuracy of WWV.

For most work, WWV and CHU can be used as the reference standard. A working transfer standard is easily made up: see Chapter 13. If this is checked about once a month and readjusted as needed, the accuracy should be adequate for most work. If greater accuracy is needed, a temperature-controlled crystal oscillator can be constructed. Alternatively, a precision signal can be obtained from the color-burst signal of any color TV receiver.

TIME STANDARDS

Time standards are easily derived from frequency, by counting cycles. The digital counter of Chapter 11 can provide 1-millisecond, 1-second, or 10-second signals. Some of the electronic clocks described in hobbyist articles and related literature can serve as a time or time/frequency standard. Don't forget the possibility of using a crystal oscillator and counters to develop a frequency between 50 and 100 Hz, and then driving a standard electric clock—50 Hz clocks are available on special order, or a divide-and-multiply scheme can be used to obtain the more common 60 Hz. This permits the clock to keep standard time rather than an arbitrary time.

VOLTAGE STANDARDS

The best inexpensive standard of voltage available to the home experimenter is two new and unused carbon−zinc dry cells connected in series. When fresh, these give a no-load terminal voltage of precisely 3.10 volts. The cells should only be used for voltage checks—if any appreciable current is drawn, this voltage changes, due to internal resistance in the cells. Also, if current is drawn for any appreciable time, chemical changes within the cell reduce the voltage. One way to obtain reasonable accuracy for a fair period of time is to use the cells only for standardizing a vacuum-tube or other electronic voltmeter, and keeping the cells in the refrigerator until they're needed again.

If you do much work requiring accurate voltages, you should consider purchase of a school-grade Weston standard cell. This can be purchased with or without a certificate of calibration. The cost of a calibrated unit is appreciable, but not exorbitant. Many high schools and colleges, power companies, and industrial companies maintain a set of standard cells and

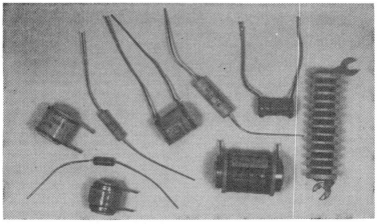

Fig. 3-2. Precision resistors of 1% and 0.5% error tolerance. The largest unit is a meter multiplier, showing typical construction of wirewound resistors. These can be adjusted to desired values. The small units are film type, and cannot be adjusted.

a voltage calibration bench. Quite often a request to them will permit checking of one of the uncalibrated standard cells...at considerable savings.

RESISTANCE STANDARDS

Precision resistors, usually to 1% tolerance of error, are often found on the surplus market at low cost. These are of two types, *film* and *wirewound* resistors. They are usually marked, sometimes with odd values, as originally required for such purposes as attenuator pads, voltage dividers, and so on. Experience with several such surplus packages shows that the resistances are likely to be close to the tolerances indicated, but cannot be fully trusted. Resistances are probably stable, and suitable for use as standards when the exact value is checked. This surplus material is very useful for working standards. Typical units found in these surplus packages are shown in Fig. 3-2.

For reference standards, the recommended procedure is to purchase one or two new 1% resistors or even one or two 0.1% resistors. These should be mounted so that there is no temptation to use them for day-to-day work. These standards could be checked to a tolerance of error of perhaps 0.01% by recourse to one of the industrial or school labs mentioned above, since most have these basic resistance standards.

It is relatively easy to adjust the resistance of the spool-type wirewound resistors shown in Fig. 3-2. It is done by removing the outer protective cover, unsoldering the outer end of the winding, and then removing turns or adding wire from another resistor. Save any wire removed, for future use.

CAPACITANCE STANDARDS

Capacitors are rarely available to the same accuracy as resistors. In addition, there is good reason to mistrust surplus capacitors, but it may be noted that the "TU" tuning units of World War II surplus contain a number of high-quality mica capacitors, including some which have been measured to within 0.1% and marked with the measured value. In spite of their age these may still be satisfactory. New 1% capacitors are not too expensive in sizes up to 0.01 μF. Units of higher precision or larger capacitance value are sometimes found in surplus, but new units are usually very expensive. Figure 3-3 shows a number of types of capacitors which could be used as working standards after the exact values are determined.

Probably the best procedure is to buy one or two good-quality mica, Mylar, or polystyrene film capacitors, to

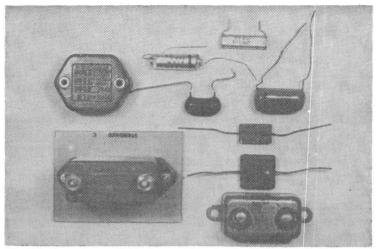

Fig. 3-3. Capacitors suitable for working standards. The two largest units are mica, intended for use in transmitters: one is mounted on a base, as a reminder that this is a standard. The others, from top to bottom, are: an oil-filled paper "bathtub"; two receiving mica types; two Mylar; and two polystyrene units.

1% guaranteed error tolerance, and to use these as your secondary standards. Dipped or molded mica capacitors can then be used for the smaller working standards. Larger capacitance units can be made up from oil-filled paper or Mylar capacitors, using two or more in parallel to secure the desired value. Industrial and power company laboratories often have good capacitance standards, which will permit calibration to within 0.1%.

INDUCTANCE STANDARDS

Inductance standards are a difficult problem, and attaining accuracy tends to be very expensive. The best procedure appears to be construction of a set of working standards, and calibration of these by use of capacitance and frequency standards. Chapter 10 shows a standard which can be used with fair accuracy if dimensions are kept as shown. For 10 and 100 millihenry (mH) standards, it is possible to add or remove turns from telephone-type toroid inductors as described in Chapter 13. If much work is done with large inductors, it is worthwhile to make up a 1-henry(1H) standard by placing a dozen of the 88 mH telephone inductors in series, removing turns from one of these to secure the 1H. Coil quality or Q is difficult to measure with accuracy, and Q standards are expensive. Figure 3-4 shows two such standards, plus two coils which can be made into transfer standards.

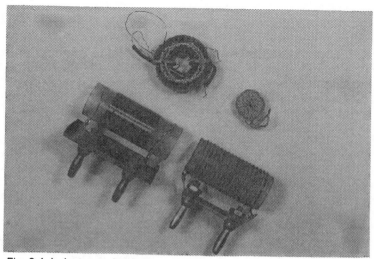

Fig. 3-4. Inductors suitable for standards. The coils on plug-in bases are for RF use, with the Q-meter of Chapter 10. The smaller toroid is also usable at RF. The larger toroid is an 88 mH telephone loading coil, useful in constructing standards from about 2 mH to 1H.

OTHER STANDARDS

A relative standard for light intensity can be set up by placing a new 60W bulb at a known distance from the work area to be illuminated. If the line voltage is adjusted to the value marked on the bulb, the light intensity is equal to the value on the carton in which the bulb is shipped, divided by 12.56636 (the number is 4π). This relation is valid only for new bulbs, since blackening of the inside of the bulb due to filament evaporation reduces the light output. A check of this light intensity can be made using a photoelectric exposure meter in accordance with its instructions.

Standards for acoustic intensity can be calibrated using the procedure based on electrical quantities described in Chapter 12. If you plan to do much acoustic work you should consider building up a small single-frequency oscillator/speaker combination. The sound intensity produced on the axis of the speaker and at a standard distance is constant if some precautions against the effect of stray sound reflection are taken. Once calibrated, this should give constant output if the battery voltage is kept at the same value.

A special standard for electric current is not usually worthwhile. The best procedure is to calibrate a good quality ammeter or milliammeter, using a stable wirewound resistor as a working standard, with reference to a calibrated voltmeter.

Standards of length of various levels of accuracy are obtainable at most hardware stores. Standards of mass can also be purchased, in the form of scale weights. Standards for other quantities may be desirable if special work is undertaken. These can be developed as suggested in the books dealing with measurements in the particular field.

PROTECTION OF STANDARDS

It is important to remember that whatever type of standard is secured and whatever calibration technique is used, the reference standards should be protected. They should be kept separate from regular equipment stock so that they will not be accidentally (or intentionally!) used. Unnecessary handling should be avoided, as should exposure to overload. This separation is also good for transfer standards, but occasional damage of such a standard is no great problem since another can be constructed, if necessary, or recalibration accomplished. Good practice is to check transfer standards several times a year and to check the reference standards against those of another laboratory whenever opportunity offers, hopefully at least once every few years. A log should be kept of such checks as a way of detecting drift in the standard.

CALIBRATION PROCEDURES

The steps in calibration involve:

- Transfer from reference to working standards
- Transfer from working standards to instruments
- Preparation of calibration data
- Checking calibration, or standardizing

In a well equipped laboratory, transfer from reference to working standards is usually done by matching or, at most, determination of differences. For example, if a 0.05 μF capacitor is to be used as a working standard, it will be compared to a 0.05 μF reference standard, probably in the form of a decade capacitor. The reason for this preference for determination of small differences is easily seen; suppose that the unknown has an actual capacity of $(0.05 + x)$ μF. If x is small—say, no more than 10% of the unknown, and is determined only to an accuracy of 10%, the error in determining the value of the standard is no more than $x/10 \times 0.1$ or 1%.

The small lab should use this matching or difference determination wherever possible, but usually will not have a large enough assortment of standards to be able to depend on it. Thus the small laboratory must place dependence on scale factor multiplication, on conversion of unknowns, and on interpolation. Scale factor multiplication means working at a known multiple or submultiple of the value being measured, such as a factor of two, five, ten, one-half, one-fifth, etc. Conversion depends on having standards for some quantity which is easy to measure, such as resistance—in essence, converting the calibration variable to this easily measured quantity. One of the more common methods of conversion is to prepare a temporary calibration in terms of angle, as measured for a variable resistance or capacitance.

Interpolation is a variation of conversion. It allows determination of values between two points at which standards are available. (If suitable precautions are taken, *zero* becomes a calibration point.) Interpolation may be done by instruments—for example, by a variable voltage divider. The RC(L) bridge described in the next chapter contains such an interpolating resistor. Several interpolating instruments are commercially available. Interpolation may also be separately performed, and for the small lab it is more likely that interpolation will be graphical, a special form of conversion is illustrated in Fig. 3-5 and reviewed further along in this book. There is a companion technique, extrapolation, or going beyond a point of calibration. The danger of use of this technique, also shown in Fig. 3-5, is that a nonlinearity may be present, such as saturation, which introduces large errors. However, used with caution, extrapolation is useful.

Checking the calibration of an instrument should be done at regular intervals. If there is no reason to suspect change, a single-point calibration check is usually satisfactory. It should be supplemented, from time to time, by a repeat of the full calibration. A good procedure is to make a single-point check every two or three months, and full calibration every one to two years. If variations are found during the single-point check, the instrument should be fully recalibrated.

Some instruments are intended to be calibrated at each use. This is usually due to difficulty in securing stability, for example, to eliminate seasonal temperature changes. It may be due to such problems as maintaining constant output of an oscillator as its frequency is varied. The Q-meter described later is to be set to proper output at each frequency at which Q measurement is made. This particular form of calibration is often called *standardization*.

Rather than attempting a general explanation of calibration procedures, let us illustrate the techniques by two

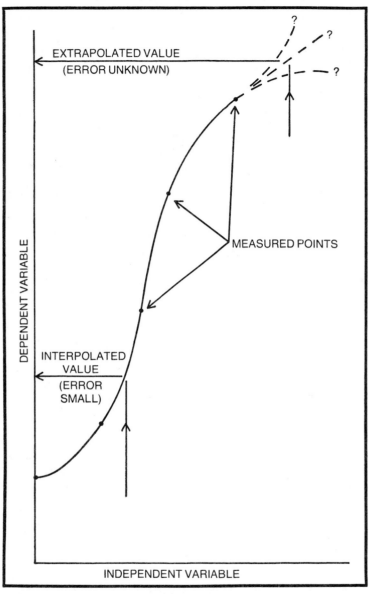

Fig. 3-5. Graphical illustration of interpolation between measured values, and extrapolation beyond a measured point. Since the trend is unknown, extrapolation may be subject to large errors.

examples. The first can illustrate the establishment of a calibration by calculation and conversion; and the second, calibration by measurement and reference to standards.

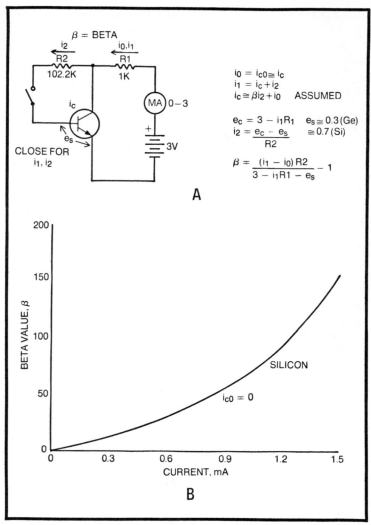

Fig. 3-6. Illustration of calibration by calculation. (A) Circuit of typical transistor checker, and derived equation of performance. (B) Calculated calibration curve, which can be marked directly on the meter face if direct reading is desired.

Calibration by Calculation

Figure 3-6 shows the basic circuit of a transistor checker used on the low current scale. Shown on the drawing are element values and assumed current flows and voltages.

To solve this circuit for the major parameter of interest, the transistor current gain or DC beta (h_{FE}), some assumptions

43

about the transistor are needed. Suppose it is assumed that the transistor internal resistances are negligible, and that the leakage current and base—emitter drop are independent of current. The equation for beta can now be calculated, as shown on the figure. A sample calibration curve obtained from this is shown in Fig. 3-6B, for $i_{C0} = 0$ and a silicon transistor.

It is instructive to study the error introduced in measurement by variations in circuit values—for example, by variation of battery voltage or resistance values. It is also instructive to investigate the effect of neglected quantities, such as transistor internal resistance, or the way the beta curve varies with small amounts of leakage current.

Those who use this type of instrument may want to make up a calibration curve or table, or even calibrate the meter dial for the transistor type most commonly used.

This example illustrates the process of determining one variable in terms of others—in this case determination of current gain by measuring current, with resistance and voltage as parameters. It is a process which is used quite often. For example, one of the instruments in this book measures Q by determining voltage, and another measures sound level similarly. Others determine the desired variable by varying another quantity—for example, determining frequency by varying a calibrated resistance.

Calibration by Measurement

For the second example, let us assume that the laboratory VTVM is to be fully calibrated, a procedure which is recommended both for training and for improving the accuracy of the instrument when used. A voltage standard of two new, type D dry cells will be needed, as described earlier. It is also assumed that an oscilloscope is available, as well as the usual assortment of laboratory parts.

The first step is preparatory and is for development of interpolation capability. To get this in simple form, we will need 10 resistors of the same value—the exact value is not important, as long as they are small compared to the 11-megohm VTVM input resistor; a value of about 1000 ohms is satisfactory. One way to pick these out of an assortment is to let the VTVM warm up thoroughly, set the zero and full-scale points, and then select 10 resistors which give the same indicated resistance readings, as closely as can be read. (Of course, if the bridge described later has already been built, it would be used for this.) If necessary, two resistors can be placed in series, or two in parallel, to get the same indicated value. When 10 such matched resistors have been selected,

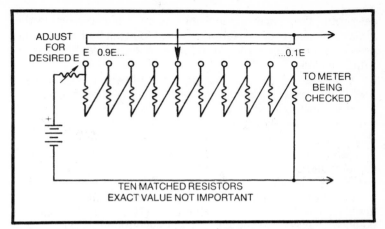

ADJUST
FOR
DESIRED E E 0.9E...   ...0.1E

TO METER
BEING
CHECKED

TEN MATCHED RESISTORS
EXACT VALUE NOT IMPORTANT

Fig. 3-7. Resistive interpolator, for checking meter calibrations. With suitable auxiliary elements, can be used to check AC and DC voltage and resistance calibrations.

they should be connected in series and then to the lab supply, with a variable resistor in series, as shown in Fig. 3-7.

To start the actual calibration, set the VTVM to an appropriate scale, typically the 0−5V scale, and measure the voltage of the two series-connected reference cells. Suppose that this gives a reading of 3.17V. Now adjust the power supply or the variable resistance until the VTVM reads the same value (3.17V) when connected to the tenth point on the resistor chain. Now the intermediate taps of the resistor chain will have a voltage which is equal to a submultiple of the voltage of the standard, or 3.10/10 times the tap number. Read and record the indicated voltage on each of the 10 points of the chain, including zero, arranging them in a table. The true and indicated voltages can be subtracted to give the error. A sample chart is shown in Table 3-2.

To make the data readily useful the figures should be plotted as a graph, in one of two forms. One method is to plot the indicated value versus the actual value, as in Fig. 3-8A. This form is most useful in initial calibration. It is easy to

Table 3-2. Example Of Calibration Chart.

Step	True Voltage	Indicated Voltage	Error
10	3.10	3.17	− 0.07
9	2.79	as read	etc.
8	2.48	as read	etc.
etc.	etc.	etc.	

determine the proper scale factor to use, and to develop the "best fit" curve of variation. Such a plot should always be prepared as a step in the preparation of a direct calibration of a dial or scale.

The other method of plotting the calibration data is more useful for correcting measurements. It is made by plotting the error against the indicated value, as shown in Fig. 3-8B. The true value of an unknown is determined by adding (algebraically) the error at the indicated reading to the indicated reading. This plot is useful because it can be read rapidly, and because it gives magnification of the error. It also permits visualization of the shape of the error curve: if a mistake has been made, the bad calibration point will almost always lie well off of a smooth curve. (However, it should be noted that error curves are not always smooth, especially if the error is small.)

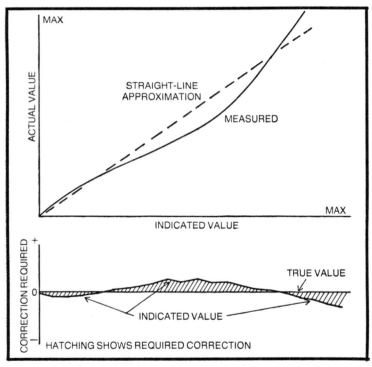

Fig. 3-8. Typical calibration curves. (A) Plot of actual value versus measured value, showing "best fit" curve and straight line approximation. Most useful in preparing a calibrated dial. (B) Plot of correction to be added to or subtracted from indicated value, versus indicated value (most useful in making measurements).

To complete the check of the 0—5V scale, and to check other scales, several methods are possible. One is to increase or decrease the voltage across the interpolating divider by adjusting the power supply or series resistor. For example, to complete the check of the 0—5V scale, adjust for 3.17V indicated on step 5, matching the voltage of the two standard cells, which gives 6.20V on the tenth resistance step. This method is satisfactory for small changes, such as two to one. However, the error in matching resistors is multiplied by the scale factor, so the error can be considerable if the multiplication factor is large.

A better way of checking the other scales is to use the RC(L) bridge of Chapter 4. Measure the resistance of each of the multipliers in the VTVM voltage-divider circuit, and calculate the value of multiplier error by ratios. The correction for the several points on the scale will be the basic correction (Fig. 3-8B) times the multiplier error.

Because of nonlinearities, a separate calibration is needed for the AC voltage scales. For this calibration a method of transferring from DC to AC measurement is needed. The oscilloscope can provide this by working at the peak of the AC curve. This is the value actually read by all inexpensive vacuum-tube voltmeters, even though their calibration is in terms of RMS voltage, 70.7% of the peak voltage (a sine wave is assumed). To use the scope as a transfer standard, first adjust the scope gain to give a convenient deflection from the dry-cell standard, say 3.1 cm. The accuracy of the calibration check will depend mostly on the accuracy of this setting. Connect the voltage divider, oscilloscope, and the VTVM to an AC supply. Measure the voltage on each step of the voltage divider with the VTVM, and record these measurements in tabular form, as in Table 3-3.

Complete the table as shown, by first converting the measured peak voltage to RMS voltage, then determining the error. Repeat this for each of the low-voltage AC scales and for the general scale. Prepare error curves and determine multiplier errors as for the DC calibration.

Table 3-3. Example of AC Calibration.

Divider Step	Peak Voltage	RMS Voltage	Meter Readings	RMS
10	3.1	2.19	2.09	+0.1
9	27.09	1.915	as read	etc.
8	24.08	etc.	etc.	
etc.	etc.			

It is possible to make such a calibration for resistance, but it is not necessary. The meter face markings are based on calculations and should be correct. It is only necessary to check the internal resistance standards. On a vacuum-tube voltmeter of the usual type, the design value of this internal standard can be found by noting the resistance value marked at the midpoint of the resistance scale. For example, in the old EICO Model 221, the midpoint value is 9.5, or 9.5Ω for the $\times 1$ scale, 95Ω in the $\times 10$ scale, etc. The value of the internal resistors can be checked by measuring the apparent resistance of a standard which gives about half-scale deflection. The error in measurement is the error of the internal standard. For example, if a 10K reference standard is measured on the $\times 1000$ scale, and the indicated resistance is 10.5K, the internal standard will be $10.5 - 10.0/10.0$ or erring by 5%. (The alternate method of checking the internal resistances is to use a resistance bridge.)

THE INITIAL CALIBRATION

The above basics apply to all calibrations. However, special problems exist when an instrument is first calibrated. One is faced with a blank dial or a blank panel, and must end up with a calibration. The solution is to introduce an arbitrary scale and perform the preliminary calibration steps with respect to this arbitrary scale. Later, the arbitrary scale is replaced with a direct calibration. Usually, the most convenient way of getting an arbitrary scale is to use a standard dial or one prepared with a piece of graph paper (Chapter 2).

The initial steps of calibration are as just described; however, the measured values are related to the dial reading in an arbitrary way. While it is possible to prepare and use a graph of this arbitrary relation, as is done for the Q-meter of Chapter 10, it is more desirable to make up a special dial—one which is direct reading.

To do this, it is first necessary to decide on the type of scale. Three major types are *linear, square law,* and *logarithmic*. Examples of each type are shown in Fig. 3-9. Inspection of the tabulated calibration values will permit selection of a suitable scale.

When the scale is selected, the next step is to prepare a graph, one axis being the arbitrary scale used initially, the second axis being the selected final scale. Commercial graph paper can be used, or you can draw a special piece using a slide rule for a scale. The calibration curve of the form of Fig. 3-8A (and 3-6B) is then drawn. From this, you can work out the

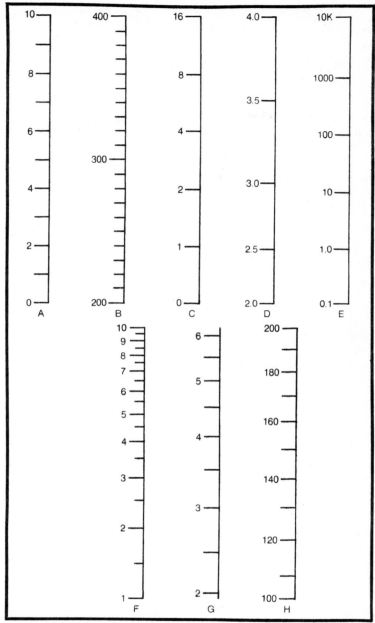

Fig. 3-9. Typical scales used in calibration:
(A) linear; (B) linear, suppressed zero; (C) square law; (D) square law, suppressed zero; (E) logarithmic, multidecade; (F) logarithmic, single decade; (G) logarithmic, half decade; and (H) logarithmic, octave scale.

scale marking and numbering plan. Usually, the numbered points should be nearly evenly spaced around the dial. This constancy of spacing is less important for the intermediate points (those which are marked but not numbered).

It is helpful in planning scale markings to inspect the scales of some commercial instruments, either physically or through catalog descriptions. This inspection will also provide useful guides as to the size of lettering, size of scale and knob, and similar factors which affect the usability and readability of the calibrated instrument.

These techniques can be applied to any type of scale dials, dial plates, meter faces, and so on. With care in planning and layout, it is possible to place several scales on the same dial, or to add a direct-reading scale to an existing arbitrary one.

ATTAINING ACCURACY

It should be evident by now that considerable painstaking detail is involved in calibration. In fact, there is almost a direct relation—the greater the care taken, the greater the accuracy. Some people find they like this work, and devote a career to the search for improved accuracy. It is a fascinating field. But for the average builder and user, this search for perfection is not needed. The important factor is to work at the selected level of accuracy for sufficient time to develop skills, then to apply these to design and use.

As with other fields, practice is necessary, and helpful.

Chapter 4

The RC(L) Bridge

Of all the instruments found in the lab, the most fundamental is the precision bridge. This is the instrument used to measure the value of the fundamental electrical quantities of resistance, capacitance, and inductance: R, C, and L. It is the principal instrument for transfer of precision in measurement from the sources maintained by the National Bureau of Standards to calibrating standards and then to working standards.

As we have just seen, attaining accuracy and precision is a difficult task requiring great care, repeated measurements, and adequate standards maintained under controlled conditions. Because of this, precision equipment is always expensive. Commonly, decreasing the error tolerance on a component by a factor of ten will increase the cost by a factor of at least four to perhaps ten or even more. Part of this is a result of the extra labor involved and part is due to the fact that only a better grade of component will have sufficient stability to maintain the improved accuracy.

The bridge described here and shown in Fig. 4-1 is intended to give reasonable accuracy in a small package at a modest cost. It should give an accuracy of some five times greater than is usually available in shops using the simple calibration techniques described. The accuracy possible with care and with access to a calibrating laboratory is even better: about 0.1% error for resistance, and 0.5% error for other quantities. For inductors and capacitors these accuracies apply only to high-Q units—the bridge has no internal provisions for measuring Q or power factor, or value of low-Q components. However, these values can be determined indirectly.

Fig. 4-1. The RC(L) bridge. Black lettering on matte aluminum finish. The 10-turn dial allows reading the unknown to 1 part in 1000. The number 8777 below the dial is the measured resistance of the precision potentiometer used.

THE BRIDGE CIRCUIT

The basic circuit of the bridge is shown in Fig. 4-2A. The heart of the bridge is the potentiometer, labeled ARM A in this diagram. It is a 10-turn precision potentiometer, linear to 0.1%, with a 10-turn dial that can be read to 1 division in 1000. The second or B arm of the bridge is a fixed resistor whose value is equal to the resistance of the A arm potentiometer. The method of adjusting to this value is described later. The C and D arms of the bridge are made up of the standard and the unknown. The relative position of these varies with the quantity being measured, as shown. This change makes the potentiometer dial read as a fraction of the value of the standard.

The complete circuit of the bridge is shown in Fig. 4-2B. The panel layout and marking of terminals corresponds to this drawing, as shown by the photograph. This particular circuit arrangement is chosen because it makes each element of the bridge available either for separate use or for connection as a specialized bridge.

COMPONENTS AND CONSTRUCTION

If maximum accuracy is desired, all components should be purchased new; all should be high-stability units of good accuracy—with a tolerance of error of $\pm 1\%$ or less. However, entirely acceptable accuracy can be obtained by purchasing a limited number of new precision components for standards and using surplus material for the remaining components.

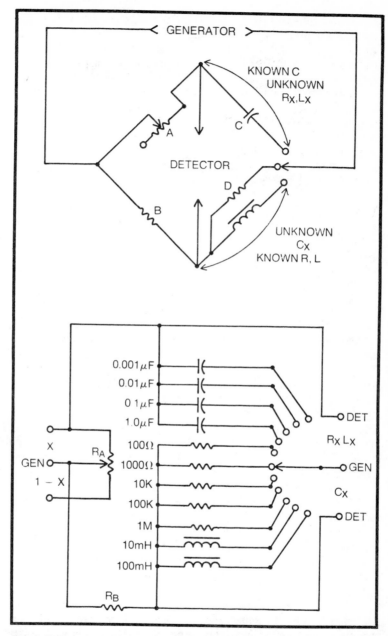

Fig. 4-2. Circuit of the RC(L) bridge. (A) Basic circuit, showing arm arrangement. (B) Complete circuit—the terminal arrangement makes all components available for special use, allows assembly of many specialized bridge forms, and allows use of external standards.

Suitable units are often listed in the ads of several of the surplus dealers who advertise in radio and electronic magazines. Precision-grade components are usable even though the exact values desired are not available, as described below.

The resistance of the key component, the potentiometer, is not critical—any value between 2K and 20K is satisfactory. If a surplus unit is used, it should be carefully cleaned and lightly lubricated with petrolatum or *Lubriplate*. These units, advertised by several surplus houses, are common components in surplus military equipment. Either the large units (about 2 inches diameter) or the miniature size (about an inch diameter) are satisfactory. If possible, obtain a unit with specified linearity.

Ten-turn dials are less common, but are often found in assembled equipment. A possible substitution is to use a standard 360° dial marked 0—100 on the potentiometer, with a small gear under this which is coupled to one of the turns counters used in tape recorders, to count the number of turns. With care in construction and choice of gear ratios, this could be made direct reading, to perhaps one part in ten thousand! In mounting either type of dial pay particular attention to dial adjustment in order to secure smooth motion that is free from backlash.

The other resistors are preferably wirewound precision resistors, but they may be metal film precision resistors. The common ceramic-form wirewound units are easily adjustable to exact value or to a special value as may be needed for R_B by carefully removing wire turns from a larger-size unit: use very fine sandpaper to remove the insulation for resoldering. The film units must be adjusted by series and parallel connection of two or more resistors.

Capacitors may be silver—mica or polystyrene in the smaller sizes, and preferably Mylar (it's capitalized because Mylar is a tradename) in the larger. Paper could be used. The only practical way of adjusting values is by parallel connection of two or more units. Kits of precision capacitors are sometimes advertised by surplus dealers at low prices. Usually these are composed of special values, sometimes unmarked.

The inductors are made from a 44 mH toroid, with approximately 11 turns removed from each winding, and with the two windings connected in parallel, start to start and finish to finish. (We discuss this in depth later.) The 100 mH unit is made by adding approximately 60 turns of AWG 30 wire to an 88 mH toroid, wound in the same direction as the original

Fig. 4-3. Internal construction of the RC(L) bridge. Standards for **R** and **C** are lead-supported, those for **L** are mounted against the case. Since this bridge is intended for low-frequency use, no internal shielding is required.

winding. Connect this and the two regular windings in series, finish of the first to start of the second, and so on.

Construction is evident from the photos of Figs. 4-3 and 4-4. The small components are supported by the switch leads and by a piece of bus wire which makes up the common lead. Inductors are mounted to the case by a bolt, being held between two pieces of plastic. No particular care has been taken to compensate for stray capacitance in the wiring. This bridge was not intended for measurement of small capacitors or inductors or for use at high frequency. Suggested panel layout is shown in Fig. 4-5, a full-size template.

Fig. 4-4. Additional view of bridge interior.

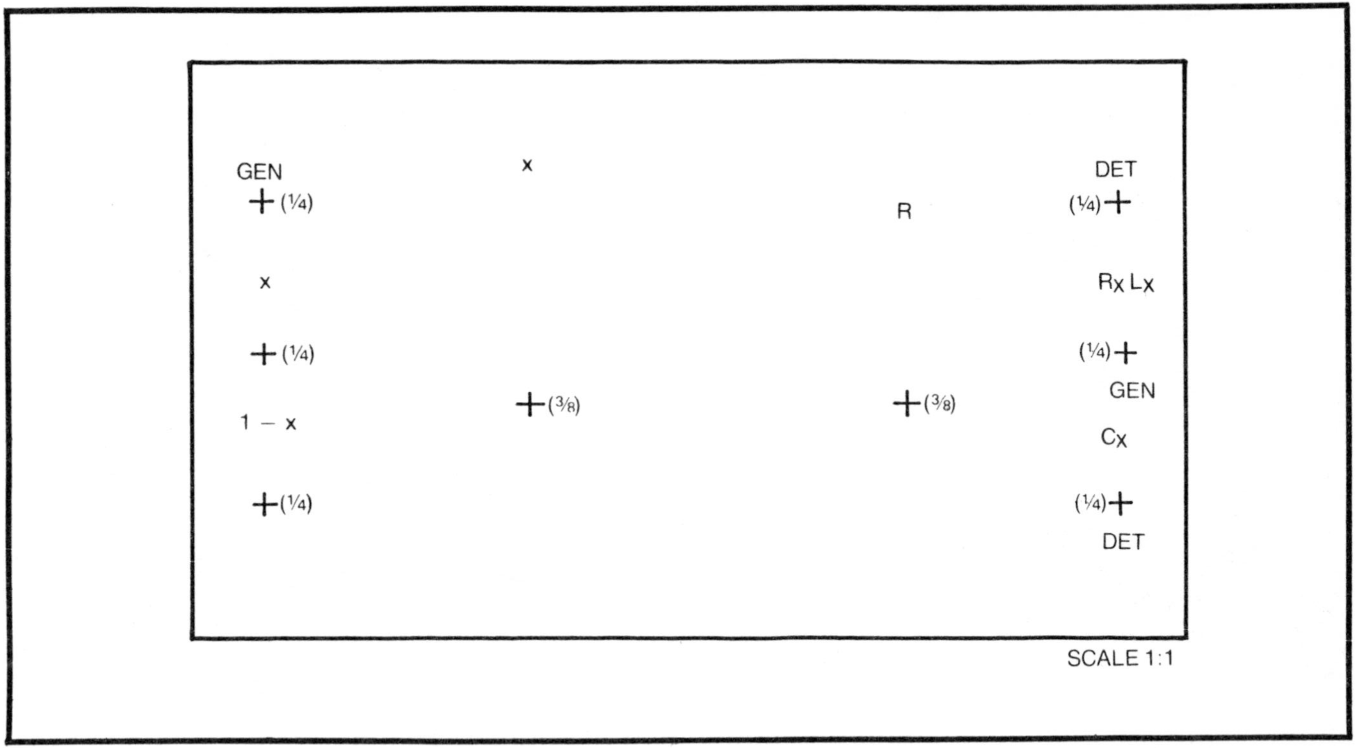

Fig. 4-5. Panel layout for bridge with components as in photograph.

SCALE 1:1

GEN
+ (¼)

×

+ (¼)

1 − x

+ (¼)

×

+ (³⁄₈)

+ (³⁄₈)

R

DET
+ (¼)

Rx Lx
+ (¼)

GEN

Cx
+ (¼)

DET

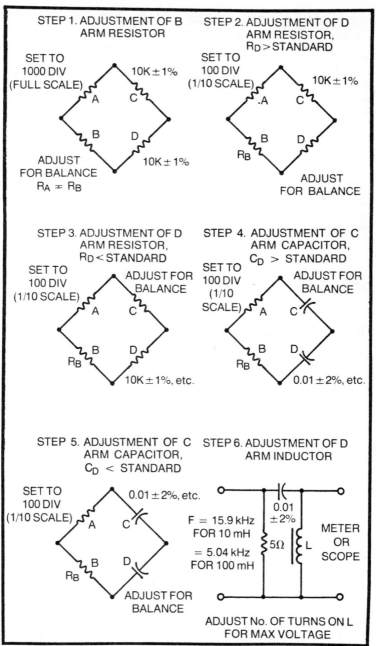

STEP 1. ADJUSTMENT OF B ARM RESISTOR

SET TO 1000 DIV (FULL SCALE)

10K ± 1%

A   C

B   D

10K ± 1%

ADJUST FOR BALANCE $R_A = R_B$

STEP 2. ADJUSTMENT OF D ARM RESISTOR, R_D > STANDARD

SET TO 100 DIV (1/10 SCALE)

10K ± 1%

A   C

B   D

R_B

ADJUST FOR BALANCE

STEP 3. ADJUSTMENT OF D ARM RESISTOR, R_D < STANDARD

SET TO 100 DIV (1/10 SCALE)

ADJUST FOR BALANCE

A   C

B   D

R_B

10K ± 1%, etc.

STEP 4. ADJUSTMENT OF C ARM CAPACITOR, C_D > STANDARD

SET TO 100 DIV (1/10 SCALE)

ADJUST FOR BALANCE

A   C

B   D

R_B

0.01 ± 2%, etc.

STEP 5. ADJUSTMENT OF C ARM CAPACITOR, C_D < STANDARD

SET TO 100 DIV (1/10 SCALE)

0.01 ± 2%, etc.

A   C

B   D

R_B

ADJUST FOR BALANCE

STEP 6. ADJUSTMENT OF D ARM INDUCTOR

F = 15.9 kHz FOR 10 mH

= 5.04 kHz FOR 100 mH

0.01 ± 2%

5Ω   L

METER OR SCOPE

ADJUST No. OF TURNS ON L FOR MAX VOLTAGE

Fig. 4-6. Steps in adjustment of arm standards, based on having two standard resistors plus a standard capacitor and a method for determination of frequency.

57

STANDARDS AND CALIBRATION STEPS

To prepare the initial standardization of the bridge, it is necessary to use a set of reference standards. A reasonable procedure is to purchase a few basic standards of fair accuracy. These are used to calibrate other units, which can be from the precision kits found in surplus, or simply high quality components.

The suggested list of basic standards is:

- 2 resistors—10K wirewound, 1 or 0.1% error tolerance
- 1 capacitor—0.01 μF, silver–mica, 1 or 2% error tolerance

One resistor and the capacitor should be mounted and labeled as reference standards: small transparent boxes with 5-way binding post terminals are convenient. These standards should *never be used for experimentation*—just for calibration. The other resistor becomes the 10K unit for arm D. (Of course, if cost is no problem, all of the standards needed for the bridge can be purchased.)

The use of these standards in adjusting the bridge arms is shown in Fig. 4-6. The first step is adjustment of the B arm. When complete, the 100K resistor is connected in the D arm and adjusted, if necessary, as shown in step 2. This is then moved to the C arm, and the 1 megohm resistor placed in the D arm and adjusted, again as shown in step 2. Connections for the 1000Ω and 100Ω resistor adjustment steps are shown in step 3. Similar operations in step 4 and 5 give the 0.1, 1.0, and 0.001 μF capacitors. For inductors, calibration is made by using the Q-meter circuit of step 6. The meter should be a high-impedance AC voltmeter or an oscilloscope.

The accuracy of the bridge is largely determined by the care taken in these calibration steps. Repeat each step several times until you have a "feel" for the process. Accuracy comes from taking pains.

USE OF THE BRIDGE

The bridge is intended for use with external generator and detector elements. Some suitable measuring combinations are given in Table 4-1.

An audio oscillator and oscilloscope combination is most useful. One reason for this is shown in Fig. 4-7. The curves show the effect of stray 60 Hz pickup for the balanced and unbalanced conditions. With headphones or a voltmeter, or with 60 Hz supply, this stray pickup can mask the null, thereby introducing error.

Table 4-1. Signal Sources and Detectors for Bridge.

Generator	Detector
1½ – 3V battery	VTVM (low scale), microammeter with series resistor which can be shorted out as balance is reached.
2.5V filament transformer	Headphones, scope, VTVM.
Audio oscillator	Headphones, scope, VTVM.

For resistors, the value measured at 60 Hz and the value of DC are almost always identical. The measuring frequency for capacitors and inductors is usually 1000 Hz; however, for inductors, better accuracy of L measurement is usually possible if the measurement is made at the frequency of maximum Q. For the common 44 and 88 mH telephone loading toroids, this is about 10 kHz.

The bridge arrangement is chosen to give direct reading of R, C and L from the potentiometer dial as a fraction of the value indicated by the selector switch. For example, if at balance the outer dial of the potentiometer reads 7, the inner dial reads 43, and the selector is at 0.01 μF, the indicated value of the unknown is 743/1000 × 0.01, or 0.00743 μF. (Note that some potentiometers will not reach 10 full turns, while others may go slightly over. The unit used in the prototype misses 10 turns by 4 small divisions, so for the most accurate work the ratio should be 743/996.)

The useful range of the bridge is limited by stray capacitance, and also by the fact that quantities being measured are not pure: inductors have resistance, etc. These factors cause the voltage at the detector terminals to be a minimum when balance is secured, but not zero, as would be obtained if separate adjustments for reactance and resistance balance were provided.

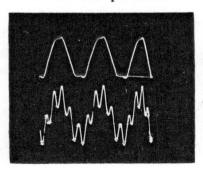

Fig. 4-7. Output voltage of bridge with arm adjusted near balance. The approximately 1000 Hz signal driving the bridge would be obscured by stray 60 Hz pickup if headphones or a meter were used for balance measurement.

An approximation to this separate balance can be obtained by driving the bridge with a square wave. With the bridge set for L or C, a good approximation to the true value of the reactance-contributing component will be obtained by adjusting for minimum imbalance of the leading edge of the square wave. Similarly, with the bridge set for R, a good approximation to the true value of the resistance component will be obtained by adjusting for minimum imbalance of the trailing edge of the square wave. While these approximations hold only for components of fairly high Q, they can be useful for low-Q components on a *relative* basis.

ADDITIONAL USES OF THE BRIDGE

The connection arrangement makes the three terminals of the precision potentiometer available for other uses. This is particularly convenient when a precision *ratio* is needed—say, the ratio of two resistances or two voltages. It also gives a precision voltage divider: for example, with 1.0V across the potentiometer, the open-circuit output can be read directly in *millivolts*.

While the bridge has no built-in provision for measuring Q or its equivalent, the bridge arms can be connected to external components to give the various specialized bridges.

Additional external standards can be made up—for example, a 1H unit made from a dozen toroids and a 10 μF capacitor. Also, calibrated variable resistances can be made that are useful for several purposes besides bridge auxiliaries.

One useful trick in measuring components larger than the internal standards of the bridge is to reverse bridge arms C and D. For example, to measure the value of a 10 μF capacitor, place a 1.0 μF external standard at the D arm, and the unknown at the C arm, with the standards selector set to a blank position. At balance, the value of the unknown is equal to the value of the standard divided by the fractional setting of the precision potentiometer. Since this is a form of extrapolation, the accuracy is reduced as compared to the normal connection.

RC(L) BRIDGE SPECIFICATIONS
- Measures resistance—10Ω to 1M (1%)
- Measures capacitance—0.0001 μF to 1.0 μF (2%)
- Measures inductance—1 to 100 mH.
- Serves as a precision variable resistor, $0-10$K
- Serves as a precision divider, adjustable to 1 part in 1000
- Bridge extendable to read D. Q.

PARTS LIST FOR RC(L) BRIDGE

A Arm: 10-turn precision potentiometer, 0.1% linearity, resistance between 2 and 20K.

B Arm: Wirewound precision resistor adjusted to resistance of A arm.

C Arm:

1 — 0.001 μF mica or polystyrene, 2%
1 — 0.01 μF mica, polystyrene, or Mylar, 2%
1 — 0.1 μF Mylar or paper, 2%
1 — 1.0 μF Mylar or paper, 2%

D Arm:

1 — 100Ω, 1%
1 — 1000Ω, 1%
1 — 10K, 1%
1 — 100K, 1%
1 — 1M, 1%
1 — 44 mH toroid
1 — 88 mH toroid

Standards:

1 — 10K, 1%
1 — 0.01 μF silver—mica 1%

Hardware:

1 — 10-turn counting dial, readable to 1 part in 1000
6 — binding posts, 5-way
1 — minibox, 2⅛ × 3 × 5¼ inches.
1 — single-pole switch, 12-position, with knob, assorted brackets, screws, wire, etc.

Chapter 5

The Audio Oscillator

Just about forty years ago a small electronic instrument organization was established on the West Coast. Originally set up to produce and market a single device, this organization has now grown to become a major supplier of all types of electronic instruments. The audio oscillator shown in Fig. 5-1 and described here is a simplified and modernized version of that original Hewlett-Packard audio oscillator. Its major characteristics are shown in tabular form at the end of this chapter.

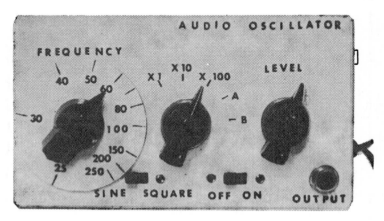

Fig. 5-1. The precision audio oscillator. The panel is finished in the plastic-on-plastic method described in Chapter 2. The jack for synchronizing is just visible on the side of the case.

THE CIRCUIT

The principle of operation of the audio oscillator is evident from Fig. 5-2. The unit is basically an amplifier with two feedback loops. The positive loop, which sets the frequency, is fed from the output of the amplifier to the noninverting input. This combination of series and parallel RC elements is derived from the Wien bridge and behaves somewhat as a low-Q tuned circuit. As shown in Fig. 5-3, the voltage across the parallel element is maximum at a frequency:

$$f_r = \frac{1}{2\pi R\,R'\,C\,C'}$$

or, for equal capacitances and equal resistances:

$$f_r = \frac{1}{2\pi RC}$$

At the same time, the phase of the signal across the parallel element is in phase with the input, also as shown. As a result, the frequency of oscillation is given by the foregoing equations as long as the phase shift in the amplifier is negligible.

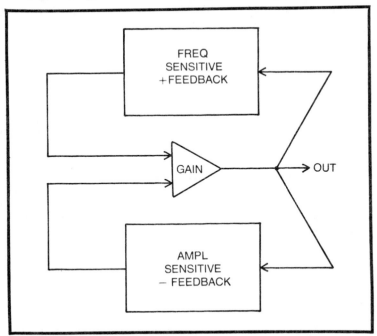

Fig. 5-2. Block diagram of a Wien-bridge oscillator. One feedback circuit establishes the frequency, the other controls the amplitude.

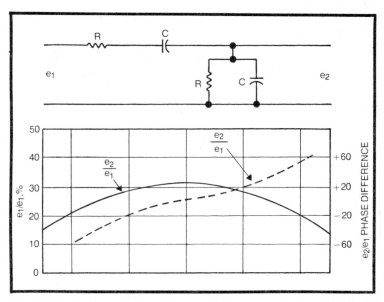

Fig. 5-3. The frequency-determining circuit. The curve of output amplitude versus frequency shows that this RC circuit has some of the characteristics of an LC tuned circuit: oscillation occurs at the frequency which gives a total phase shift (around the loop) of 360°, nominally at the same frequency as the 180° shift point of the RC circuit.

The second feedback circuit, which establishes the amplitude, goes from the output of the amplifier to the inverting input. It is really a voltage divider, composed of a fixed resistor plus a "variable" resistor (which is actually a small incandescent lamp). The resistance of the lamp changes with signal amplitude, controlling the negative feedback and thereby the amplitude of oscillation. By using heavy feedback, a sine wave of good purity is secured at the output.

The complete circuit of the oscillator is shown in Fig. 5-4. As compared to the original design, two basic changes have been made. For one thing, the active amplifying elements have been replaced by a single integrated circuit, a 741 operational amplifier. This is the major element of modernization. Additionally, the large variable capacitors used for varying frequency in the original design have been replaced by fixed capacitors, and variable resistors are used to secure tuning. This change is the major factor in attaining the miniature size of the oscillator.

In theory at least, the frequency would go to infinity if the tuning resistances were allowed to reach zero. To restrict the tuning range to a reasonable value, a fixed resistor is placed in

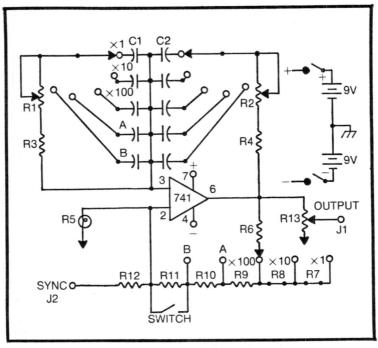

Fig. 5-4. The circuit of the oscillator. Three basic frequency ranges are provided, plus two extended ranges. The feedback can be adjusted for each range to secure best waveform. Many types of op-amp are usable.

series with each variable resistor. This resistor is chosen to give a tuning range of just over 10:1. Capacitors C1 and C2 (Table 5-1) are switched to give additional 10:1 ranges.

The oscillator includes a couple of plus features. It is synchronizable so that it can be locked to an external frequency. This also makes the instrument into a tracking filter. When synchronized, the unit can give a signal which can be varied in phase with respect to the phase of the locking

Table 5-1. Frequency-Determining Components.

BAND	FREQUENCY RANGE	C1, C2	FEEDBACK*
×1	25 – 250 Hz	1.0 µF	R6 — 200 Ω
×10	250—2500 Hz	0.1 µF	R7 — 0
×100	2.5 – 25 Hz	0.01 µF	R8 — 0
A	25 – 40 kHz	0.0010.001 µF	R9 — 100 Ω
B	25 – 40 kHz	0.0003 µF	R10 — 390 Ω
*Select for best sine wave shape.			

66

signal. The tracking filter feature can be used for detection of small phase and frequency differences, or to dig narrow-phase-shift signals out of the noise.

In addition, a simple switching circuit is provided to give square-wave output instead of the normal sine wave. This square-wave operation is secured by changing the amount of negative feedback, essentially reducing it to zero, which causes the amplifier to saturate, thereby producing a clipped (square) waveform. This simple technique does introduce a small calibration problem, since there is also an associated change in phase, which causes the frequency to change from that which would be obtained with sine-wave conditions. The frequency of the square wave will be one-third that of the sinewave.

In Fig. 5-4, five bands are provided, labeled ×1, ×10, ×100, A, and B. The corresponding frequency ranges for bands A and B are not exact multiplies of the lower bands, even when the frequency-establishing capacitors have the same 10:1 ratio (see Table 5-1). The reason for this lies in the phase-shift characteristics of the integrated circuit, which becomes appreciable at high audio frequencies. This characteristic also limits the maximum frequency of oscillation of the amplifier. The upper band calibrations and upper frequency limits will be determined by the particular integrated circuit in the amplifier.

With the 741 op-amp specified, good sine-wave performance is obtained over the range of 25 Hz to 25 kHz. (Feedback adjustment is provided for each range to allow optimization of operating conditions.) Square-wave performance is good over the range from about 6 Hz to 10 kHz. Maximum frequency of operation is about 40 kHz. All upper limits can be increased by using an op-amp capable of higher frequency operation, such as the LM301A, MC1454, etc.

None of the components in this unit are critical, although good-quality units should be used for the frequency-determining elements. The A-B variable resistors specified in the parts list are often found in surplus. If not available, wirewound types are satisfactory. If several units are available, select the one having the best "tracking" of resistance between R1 and R2 (same value of resistance in both arms at all angle settings), and low backlash. Good tracking is actually more important than exact value of resistance, which could be several times greater without affecting operation.

For the frequency-determining capacitors, it is suggested that Mylar be used for the large capacitances and mica for the smaller ones. Paper, polystyrene, and ceramic units are usable.

For good calibration accuracy these capacitors should be selected to have exact 10:1 multiples for the first three bands. The other two bands are so greatly influenced by amplifier feedback that maintenance of exact capacitance ratios is not worthwhile. However, for all bands the values of the capacitors in the series arm and the parallel arm should be identical. It is suggested that the 0.01 μF capacitors be selected by matching, and that the 0.1 and 1.0 μF capacitor values be established by paralleling with small capacitors as needed. The mounting pads for the capacitors are laid out with this in mind.

CONSTRUCTION

This unit uses a combination of PC and point-to-point wiring techniques. Because the printed circuit is very simple, and uses soldering pads of good size, it makes a good first project. The general layout of the unit can be seen from Fig. 5-1, which shows the general appearance, and from Fig. 5-5, which shows the interior arrangement. The band switch is at the center and is flanked on one side by the dual potentiometer (which sets the operating frequency) and on the other side by the output control potentiometer. The power and function switches are mounted on the top panel, as is the output jack.

Fig. 5-5. General construction of the oscillator. Most of the components are on the PC board, mounted at right angles to the panel. Feedback resistors are lead-supported from the range switch. There is room in the case for two 9V batteries.

The synchronizing jack is mounted on one side panel, a matter of convenience only. Note that the band switch in this prototype is surplus, with switch sections jumpered to give the desired connection pattern. Of course, a normal two-pole five-position switch can be used instead. Note also that the feedback resistors are mounted on this switch. This is done to simplify adjustment for best sine wave, if it should be necessary.

The drilling and mounting pattern for the integrated circuit board is shown in Fig. 5-6, and the masking pattern shown in Fig. 5-7. The method of forming the leads on the integrated circuit is shown in Fig. 2-8 (Chapter 2). These wide-spaced leads allow good-sized soldering pads on the board and are a good safety feature, since the probability of shorting from one lead to the next is reduced by the wide spacing. Where space permits, this lead form is recommended for all projects.

Drill the chassis, following Fig. 5-8. Make up the panel labeling, following the selected method described in the chapter on construction hints. Mount the panel components. Make the PC board components, and make a resistance check for possible errors.

The final assembly, after wiring of the board, is straightforward. Point-to-point wiring is used, with the wire routed directly from the switch to the PC board for the frequency-determining elements, and with it tucked along a chassis corner or wall for other elements. Figure 5-5 shows the general layout of the wiring.

In this particular unit, the battery leads were brought to the outside of the case to allow external batteries to be snapped on when the oscillator is needed. If it is to be used very often, the batteries should be mounted inside the case; there is ample room for such mounting in the area opposite the PC board. Battery-eliminator power supplies can be used instead of batteries, but there may be a tendency for the oscillator to synchronize to the power-line frequency or its multiples when tuned close to these frequencies. This particular unit was found to be free of this synchronization to the power line, unless a 60 Hz signal is injected into the synchronizing jack. If it occurs, it is due to stray capacitive coupling to the amplifier input.

CHECKING AND ADJUSTING

After completing the wiring, make a point-to-point resistance check, using the schematic of Fig. 5-4 as a guide.

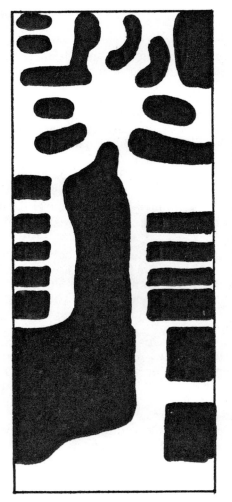

Fig. 5-6. Printed circuit layout, scale 1:1. Because of the large spacings, this makes a good initial PC project. (Original drawn freehand with ink-resist pen.)

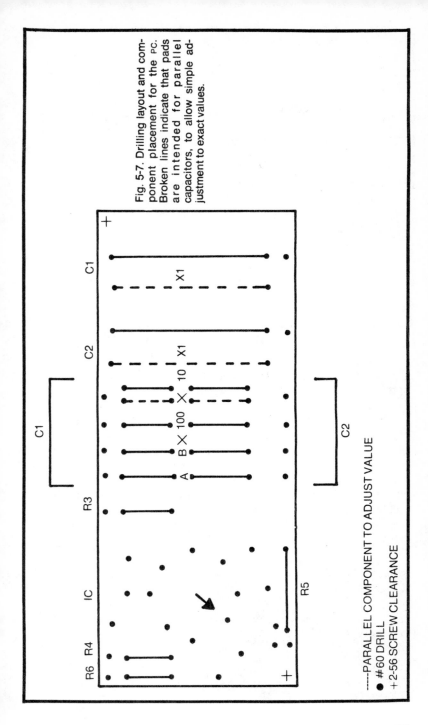

Fig. 5-7. Drilling layout and component placement for the PC. Broken lines indicate that pads are intended for parallel capacitors, to allow simple adjustment to exact values.

----- PARALLEL COMPONENT TO ADJUST VALUE
● #60 DRILL
+ 2-56 SCREW CLEARANCE

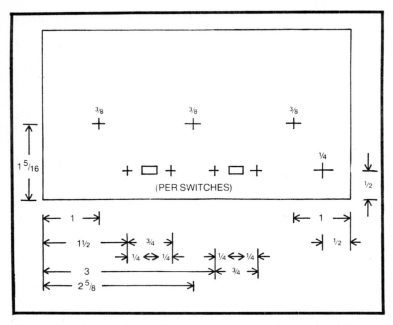

Fig. 5-8. Panel drilling layout. The switch layout in particular should be adjusted for the switches used, as there is variation in size between lots.

The most important item to check for is the polarity of the batteries—the 741 op-amp is a rugged unit but not all such units can withstand polarity reversal.

When satisfied that the unit is correctly wired, connect the batteries and turn the unit on (set it for sine-wave output). Observe the output on an oscilloscope: it should be a good sine wave on the three lower frequency bands, similar to the wave in Fig. 5-9. On the upper frequencies, the wave will become nearly triangular with rounded tips as shown in Fig. 5-10—the change is due to the slewing-rate limit of the op-amp. Oscillation should continue to a frequency of 25−50 kHz, as established by the particular op-amp.

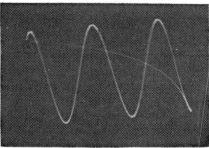

Fig. 5-9. Oscilloscope photo of sine-wave output, typical of the three lower frequency bands. Distortion is less than 1%.

Fig. 5-10. As frequency increases, the sine-wave output becomes nearly triangular, with rounded tips. The change is due to slewing speed limitations of the op-amp.

If operation is different from this, check the DC voltages first. If there is no signal, check the positive feedback circuit; if the output is not sinusoidal, check the negative feedback chain. The op-amp can be checked by touching the inverting and noninverting inputs with an ohmmeter test lead, which should cause the output to shift.

If the waveform is fairly good but not perfect, adjust the value of the feedback resistors until a good sine wave is obtained. These should be chosen to give a peak-to-peak sine-wave output equal to half the peak-to-peak square-wave output. The distortion becomes less as the feedback is increased (R smaller). However, one of the tricks of lamp-stabilized RC oscillators can appear. This is a cyclic variation in output amplitude following a sudden change in frequency and lasting an appreciable time. The *50% of square wave output* is a compromise between this and distortion.

It is not unlikely that one sine peak or the other will be flattened, as shown in Fig. 5-11. This can be caused by unequal battery voltages, or by the op-amp. One cure is to connect a variable resistor between one of the balance inputs of the op-amp and ground (pins 1 or 5 to ground). Adjust this for symmetrical operation, then wire in a fixed resistor of equal value.

When switched to square-wave operation, the output amplitude should increase, to about 50% of the total battery voltage. At low frequencies the wave should be very square with no overshoot or ringing, as shown in Fig. 5-12. As

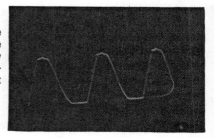

Fig. 5-11. Distorted sine-wave output. Such distortion may be due to insufficient feedback, low battery voltage, high internal battery resistance, or high offset voltage in the op-amp.

73

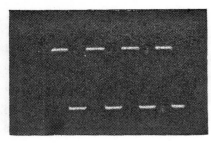

Fig. 5-12. Low-frequency square-wave output. Rise time is approximately 20 μsec, determined by the op-amp.

frequency is increased the sides start to slope but the top remains flat, as shown in Fig. 5-13. Finally, one of the tops will become rounded; the amplitude will then decrease rapidly as the frequency is further increased.

Even though a scope is not available, it is possible to get the unit into operation using a separate amplifier and an AC voltmeter. Use the amplifier as a signal tracer in getting the unit into operation, and use the voltmeter in setting the feedback. You *can* do this by ear: almost everyone can hear the point of peak-flattening, where distortion starts to be serious.

CALIBRATION

The easiest way to calibrate this unit is to use the power line as the reference frequency, and determine the oscillator frequency by establishing Lissajous figures on an oscilloscope. Calibration of the first (×1) band is easily accomplished by this method. This same calibration will be good for the ×10 and the ×100 band if the component values have been selected carefully. The same calibration technique can be used for the square-wave frequencies.

The general technique of calibration and dial preparation is covered in Chapters 2 and 3. The specific calibration frequencies most usable are shown in Table 5-2, which gives the frequency at which a stable oscilloscope figure is obtained, the number of crossings of the horizontal and vertical axis which are associated with the frequency, and the number of

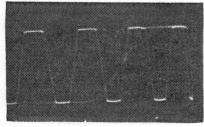

Fig. 5-13. At high frequencies the rise time becomes noticeable. If peak rounding is appreciable it is likely that one battery has high internal resistance.

Table 5-2. Calibration Points, 60 Hz Reference Lissajous Figure.

DIAL ANGLE	FREQUENCY, Hz	HORIZ CROSSINGS	VERT CROSSINGS	PERIOD, μSEC
	25.5	7	3	39216
	30	2	1	33333
	36	5	3	27778
	40	3	2	25000
	48	5	4	20833
	60	1	1	16666
	80	3	4	12500
	100	3	5	10000
	120	1	2	8333
	150	2	5	6667
	180	1	3	5555
	240	1	4	4167

microseconds for one complete cycle. The blank column in the table can be used to record the angle of potentiometer rotation which corresponds to the indicated frequency. These angles are converted to the dial calibration points by the method outlined in the chapter on standards and their use.

For the two upper bands, an approximate calibration can be prepared by using the oscilloscope as a transfer standard. However, it is suggested that calibration for these bands be delayed until the digital frequency meter of a later chapter is constructed. With it, preparation of complete calibration is simple and rapid.

The prototype unit was made direct reading for sine waves only and for the three bands, ×1, ×10, and ×100. Calibration for the square-wave frequencies and for the other bands is maintained in a separate notebook of calibrations. If desired, the square-wave calibration can be marked on the tuning dial—say, by using a different color of marking or by using concentric scales.

If a scope is not available, it should be possible to do a satisfactory job of calibration using a loudspeaker and a piano—musical training is a great help here. A good ear should be able to pick out the major chords if the 60 Hz power line is used on one speaker and the oscillator output on another.

USE OF THE INSTRUMENT

There is one precaution to observe in use: Since the output is directly coupled to the op-amp, a blocking capacitor must be placed between the oscillator and the external circuit if it carries voltage or if it could be affected by the offset voltage of the op-amp. In addition, there is the usual matter of output loading: this oscillator is designed for medium- to high-impedance loads. If it is necessary to drive a low-impedance circuit, either use a matching transformer or put an isolating amplifier between the oscillator and the load.

Other than for these factors, most laboratory uses amount to connection of a power supply, turning the oscillator on, and setting it to frequency and level. Its simplicity is an advantage.

If a synchronized oscillator or tracking-filter operation is needed, the synchronizing signal should be fed into the synchronizing jack. The amplitude of this signal should be adjusted to give satisfactory synchronization, starting with as small a signal as possible. As long as synchronization is maintained, the frequency control dial becomes a phase-shift dial. The phase shift is zero when the natural frequency of the oscillator is precisely at the synchronizing frequency. The relative phase can be varied by almost $\pm 90°$ before synchronization is lost, the exact limit varying with the amount of synchronizing signal.

AUDIO OSCILLATOR SPECIFICATIONS

- Sine-wave output: 25 Hz to 25 kHz, $4V_{P-P}$ output
- High-impedance output
- Distortion $<2\%$
- Synchronizable as a tracking filter
- Synchronizable as a phase adjuster
- Square-wave output: 6 Hz to 10 kHz, $8V_{P-P}$ output

PARTS LIST FOR AUDIO OSCILLATOR

R1, R2	— Dual 5K potentiometer, A-B
R3	— 470Ω, $\frac{1}{2}$W
R4	— 470Ω, $\frac{1}{2}$W
R5	— 12V, 70 mA lamp bulb
R6–R10	— see Table 5-1
R11	— 6.8K, $\frac{1}{4}$W
R12	— 12K, $\frac{1}{2}$W
R13	— 7K potentiometer, A-B
C1,C2	— See Table 5-1

Chapter 6

Three Useful Amplifiers

In measurements, just as in general electronic work, it is often found that a signal is too small to be usable—that it must be beefed up by use of an amplifier. This increase in signal level can be an important adjunct to securing measurement accuracy. For example, if a signal is equal in strength to an instrument noise level, a reduction in signal level by an order of magnitude—a factor of ten—will give a measurement indication of only a two-to-one reduction, a large error. Then, too, amplifiers always seem to be needed for general laboratory and experimental work, as part of a setup, to provide isolation, or perhaps as an aid to troubleshooting when an experiment doesn't work on the first trial.

The following describes the construction of three small amplifiers. One of these is for audio, one is for video, and one is for RF. The audio amplifier is laid out so that it can also serve as a signal tracer. The video amplifier has sufficient range to allow use over the entire video band and much of the high-frequency range. The RF amplifier is useful into the VHF range and, with proper transistors, is sufficiently low noise to allow use as a receiver preamplifier.

The audio amplifier requires no PC work, just assembly. The others are built on small PC boards. None of the amplifiers require calibration, although the audio amplifier is more useful if its gain control is calibrated in relative decibels. The usefulness of all amplifiers for measurement is increased if their gain and response curves are measured and recorded.

AUDIO AMPLIFIER

The basis of the audio amplifier, Fig. 6-1, is a small preassembled amplifier circuit board, sold by Lafayette,

Fig. 6-1. The speaker grille of this audio amplifier is made from perforated metal sheet, and could be screen wire. Input and output terminals may be added as needed to fit the commonly used plugs. Calibration of the gain control would increase usefulness.

Radio Shack, and other companies. Several different types of these are available, the most useful being a 4-transistor type having an output of about 100 mW and requiring an input of 300 μV for full output at maximum gain setting. The unit is transformer-coupled, but the frequency response is quite good—the curve is flat to better than 20 kHz, and useful response is obtained to about 50 kHz. Saturation characteristics of the amplifier are satisfactory.

Requirements

A block diagram of the amplifier is shown in Fig. 6-2. The required external components are gain control with switch, a speaker, and a set of input and output terminals.

To increase the usefulness of the amplifier several parallel sets of input terminals should be provided. In the prototype only one type of output terminal was installed, but more can be added if desired. Two different input circuits are available, one for maximum gain, with the level controllable by the setting of the gain control, the second at a lower level and at fixed gain. The second input makes use of a connection on the amplifier originally intended for tone control. In common with all units of the miniature laboratory series, the amplifier is operated from an external battery supply. There is ample space in the case if internal mounting of the battery is desired.

The panel layout used in the prototype is evident from Fig. 6-1, and the general layout and construction are shown by Fig. 6-3. No special precautions in construction or layout are

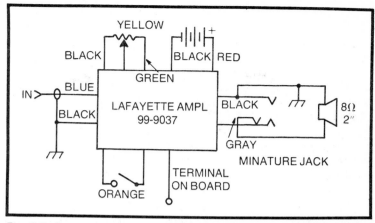

Fig. 6-2. Connection diagram of the audio amplifier. Units from other sources will vary in details.

needed. The prototype finish is bare metal, scoured with steel wool to give a satin finish and protected with a coat of transparent lacquer. Lettering is applied over the lacquer and a second layer of lacquer applied to protect the lettering.

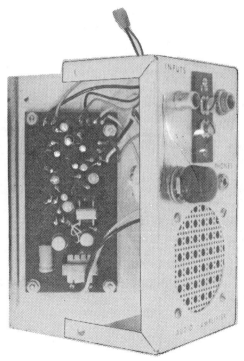

Fig. 6-3. Internal arrangement of the audio amplifier. There is ample space in the case for a 9V battery.

Calibration

Calibration of the amplifier is recommended. The procedure is to set the gain to maximum and apply a known input, say 1 mV, from an audio oscillator or from a 60 Hz transformer. This may be obtained by feeding a signal at a higher level to the voltage divider of the RC(L) bridge and setting the precision potentiometer to give the desired division ratio.

You can insure against overloading by viewing the output on an oscilloscope or by determining that the output varies in proportion to the input with no saturation. The amplifier input is then adjusted to give constant output for various angle settings of the gain control, using a meter or an oscilloscope at the output. When the gain-versus-angle measurements have been made and recorded, a calibrated dial can be prepared in accordance with the chapter on construction techniques. It is also desirable to measure the gain from the fixed input to the output, and to determine the overall frequency response. The response is best measured by determining the input for constant output over the frequency range from about 20 Hz to 50 kHz.

The values for all measurements can be kept in notebooks; however, it is sometimes convenient to mark the essential data on the front panel—for example, by marking the input-to-output gain at the fixed terminal or by putting the limits of frequency response directly under the title of the instrument.

Incidentally, if the sensitivity of these amplifiers is below the value given by the instructions, it is a good idea to check the electrolytic capacitors. Some of these units may be several years old, having been in storage for this period. Also, it may be desirable to change the output transistors to a more modern type. If this is done, the amplifier can be used on a 12V as well as a 9V supply, and more output can be obtained before saturation. With modern transistors the output saturation point will be determined by the transformers rather than by the transistors.

VIDEO AMPLIFIER

The video amplifier (shown in Chapter 2, Fig. 2-2) was designed to provide a specific gain and to operate over a wide band of frequencies. For the prototype, the voltage gain was set to a value of $\times 10$, or 20 dB, which is especially convenient for measurement work. The gain can be set to other values during construction, if desired: gains from 10 to 30 dB are obtainable. Strictly speaking, the specified gain will be obtained only when the amplifier is used at the load for which it is calibrated. However, the gain variation will be reasonably small over a wide range of loads. This is due to the use of inverse feedback, which sets the gain and stabilizes it.

The frequency response is the video range from 30 Hz to a minimum of 5 MHz—and with good transistors the frequency range is even better. In the prototype, using inexpensive audio transistors, a frequency response of 5 Hz to 15 MHz (at the 3 dB points) was obtained.

The circuit diagram of the video amplifier is shown in Fig. 6-4. It is based on a design by Texas Instruments, originally prepared for germanium PNP transistors. In this design, NPN transistors are used, with some of the component values modified.

Two feedback loops are provided in the amplifier. One is from the output transistor emitter to the input transistor base, via a resistor. This DC feedback loop provides thermal stability and sets the idling current of the transistors. The second feedback loop is from the output collector to the input emitter

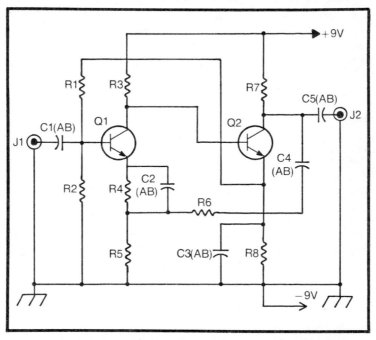

Fig. 6-4. Circuit diagram of video amplifier. One feedback path sets the idling conditions, the second independently sets the AC gain. This unit does not use a power switch or a bypass switch, which could be added.

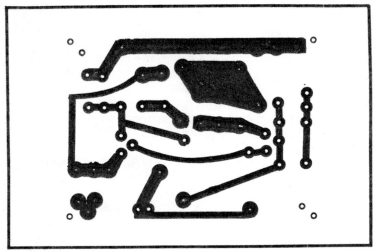

Fig. 6-5. Board layout for the video amplifier—made with "Easy-Etch" transfers. (Original board drawn freehand with an ink-resist pen.)

and is AC-coupled. This loop establishes the amplifier gain by the ratio of R6/R5 + R6. The two stages are direct-coupled. Input and output coupling is capacitive. The coupling and bypass capacitors consist of a relatively small Mylar capacitor in parallel with a large electrolytic, the combination providing good low- and high-frequency response.

The board layout for the amplifier is shown in Fig. 6-5; component placement in Fig. 6-6. No special precautions were taken with layout. The prototype uses sockets for the transistors, but these may be omitted with a small increase in high-frequency response. Operation is from a single 9V battery. There is room in the case for the battery if desired. General construction of the amplifier is shown in Fig. 6-7. The case is a nominal $1 \times 2 \times 3$ inch aluminum box. In the prototype, a hammertone finish was selected. Lettering for the prototype was by white pressure-sensitive letters protected by a coat of clear lacquer.

Initial setup will require adjustment of two resistors. R3 provides the DC feedback, and should be adjusted to give DC voltages close to the values shown on the diagram. The value (see parts list, end of chapter) was found to give satisfactory idling current with over half of a selection of surplus plastic audio transistors, the type found in the "50 for $2" assortments.

The AC gain is set by R6. The values shown in the parts list should give very nearly 20 dB gain with transistors having h_{FE} of about 50. The value of this resistor can be varied to give

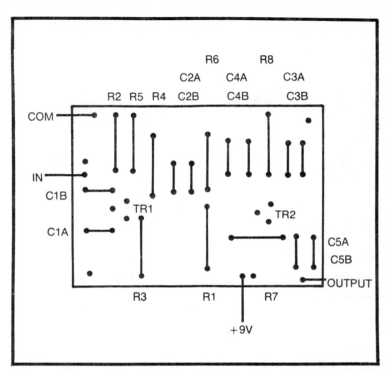

Fig. 6-6. Drilling template, component layout, and lead connection of the video amplifier. All capacitors are formed using an electrolytic and a disc ceramic unit in parallel.

Fig. 6-7. Internal view of video amplifier. The RF amplifier is similar in construction.

other values of gain, but don't go for gain less than 10 dB or more than about 30 dB: high gains require low feedback and high-gain transistors and tend to be unstable; lower gains may introduce problems with oscillation. A convenient way of setting the gain is to feed about 0.1V of high-frequency audio (20 to 30 kHz) to the input, using a scope with a 10:1 switch as a voltmeter at the input and output: adjust until the gain is the value desired.

If a very low-frequency response is needed, the size of the electrolytics should be increased to 100 μF or more. The upper frequency response can be extended considerably by using RF transistors instead of the audio types of the prototype. It is recommended that the upper and lower half-power (3 dB) points be determined by measurement and recorded as a guide to future use or in case repairs are needed.

In the prototype, RCA-type phono jacks were used for input and output connectors. It may be convenient to place an additional set of connectors in parallel with these, using the same type as found on the lab oscilloscope.

RF AMPLIFIER

The RF amplifier is intended to be useful to frequencies above the FM broadcast band. It is a simple design: a common-emitter amplifier for gain, followed by an emitter follower for isolation and impedance matching.

The exact gain and frequency response depend on the transistors used. Using low-cost RF transistors, gains of the order of 30 dB at 10 MHz and 10 dB or so at the FM broadcast

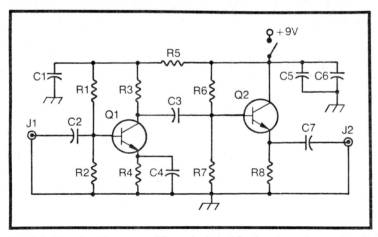

Fig. 6-8. Circuit of the RF amplifier—designed for low-impedance circuits such as transmission lines.

band can be expected. The low-frequency response varies with the output impedance, but extends to about 100 kHz with a 50Ω load.

The circuit diagram of the RF amplifier is shown in Fig. 6-8. Mylar capacitors are used for coupling, but could be replaced with polystyrene, mica, or even ceramic. Smaller values than those could be used if low-frequency response is not important. Parallel Mylar and electrolytic capacitors are used in the supply circuit, to insure low-frequency stability.

The board layout of this amplifier is shown in Fig. 6-9. The relatively large areas of ground plane are a way of reducing stray coupling. Parts placement is shown in Fig. 6-10. The panel layout is evident from the photo of Fig. 2-3, and general construction from Fig. 6-7. Note that in the prototype the power switch also is used to feed the input signal to the output when the amplifier is off. This cross feed should be omitted if the amplifier is to be used in a high-impedance circuit as the stray coupling through the switch may be sufficient to cause oscillation problems.

As with the other amplifiers, the upper and lower half-power points should be measured and recorded for future use.

USES OF THE AMPLIFIERS

The areas of application for these amplifiers follow from their names. The audio amplifier can serve for both laboratory and general servicing work. Its usefulness is greatest if it is

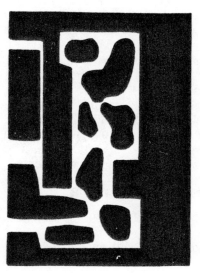

Fig. 6-9. Layout of RF amplifier PC board (drawn freehand with India ink). This shows the etched side of board. To verify component placement, trace outline on blank sheet, then turn sheet over. Lay tracing of Fig. 6-10 over it and hold up to light. Component placement must register with PC mask.

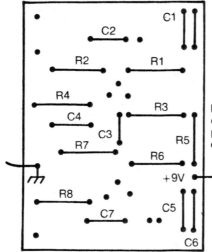

Fig. 6-10. Drilling template, component layout, and lead placement for component side of RF amplifier board.

equipped with external probes. A low-capacitance probe and an RF probe are the most useful. These probes may be the ones used with the laboratory oscilloscope, providing one of the input connectors has been chosen to correspond with that on the oscilloscope. Used with an RF probe and an auto radio antenna, the unit has been used for noise tracing, to find the noise source interfering with shortwave reception.

If low-noise transistors are used, the noise figure of the RF amplifier should be better than many receiver front ends. It can be used as a wideband amplifier, but will probably require some form of RF selectivity ahead of it, or cross modulation may develop. The selectivity can take the form of a bandpass filter or a tuned circuit.

The video amplifier is a useful adjunct to the AC section of the vacuum-tube voltmeter, especially if the gain is set to precisely 10:1. With the usual type of VTVM, this will give a full-scale AC range of 300 to 500 mV. If some precautions are taken in grounding and shielding of the leads, it should be possible to cascade two of these amplifiers to give gains of up to 100 for VTVM sensitivities of 30−50 mV full-scale. The amplifier is equally useful for AC oscilloscope measurements.

PARTS LIST FOR AUDIO AMPLIFIER

1 — Audio amplifier module, equal to Lafayette 99-90375, 330 mW output, 300 μV input
1 — Volume control, 10K (audio taper), with switch and knob
1 — Speaker, 2″, 8Ω

1 — Battery leads
1 — Aluminum chassis, 2⅛ × 3 × 5¼ inches
Input and output terminals, assorted nuts, bolts, spacers, etc.

PARTS LIST FOR VIDEO AMPLIFIER

R1 — 22K All resistors ½W
R2 — 5600Ω
R3 — 12K
R4 — 220Ω
R5 — 82Ω
R6 — 10K
R7 — 1000Ω
R8 — 1000Ω
C1−C5(A) — 100 pF ceramic
C1−C5(B) — 100 μF, 15V
Q1, 2 — NPN, 200 mW, h_{FE} = 30 min, f_t = 150 MHz min

PARTS LIST FOR RF AMPLIFIER

R1 — 22K   All resistors ½W
R2 — 10K
R3 — 1500Ω
R4 — 1000Ω
R5 — 100Ω
R6 — 4700Ω
R7 — 4700Ω
R8 — 270Ω
C1−5 — 0.1 μF Mylar capacitor
C6 — 10 μF, 15V electrolytic
C7 — 0.1 μF Mylar capacitor
Q1, 2 — NPN, 300 mW, h_{FE} = 30 min, f_t = 200 MHz min
(RF: 2N3694, etc.)

Chapter 7

The Step-Gain Amplifier

In many laboratory problems a range of gain values is needed where the gain is at once known, easily resettable, and quickly readable. Examples of such problems are calibration of an amplifier, comparison of two signal levels, or determination of the degree of improvement in a project. The step-gain amplifier is intended to satisfy these needs and to serve as a useful general-purpose amplifier for DC, audio, and the ultrasonic range. Specifically, it is intended to provide gains from −40 to +40 dB, adjustable in 2 dB increments.

THEORY OF OPERATION

The step-gain amplifier, Fig. 7-1, is an operational amplifier with feedback. The basic circuit of this amplifier is

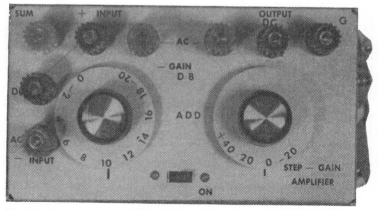

Fig. 7-1. The step-gain amplifier. The gain is the algebraic sum of the setting of the two dials. Panel finished with black lettering on white background. Dials are homemade, as described in Chapter 2.

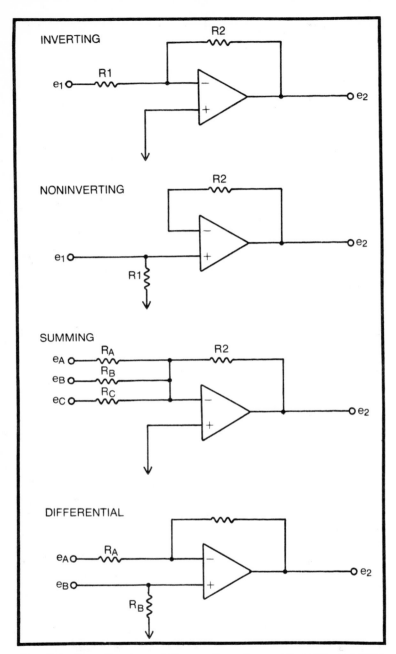

Fig. 7-2. Operational amplifier configurations: (A) basic inverting amplifier; (B) basic noninverting amplifier; (C) summing amplifier, inverting; (D) differential amplifier, combining (A) and (B).

shown in Fig. 7-2A. Assuming that the amplifier gain is very large, the gain of the feedback circuit can be written as:

$$e_2/e_1 = R2/R1$$

This can also be written:

$$\log e_2 - \log e_1 = \log R2 - \log R1 = \text{gain, dB}$$

We can thus set the gain of the amplifier by choice of R2 or R1. It is convenient to vary one of these resistances to give 10:1 steps in voltage gain, or 20 dB, and to vary the other for intermediate steps, a step of 2 dB being used in the prototype. Note that when R2 is equal to R1, the gain is unity (0 dB), the reference point for calibration.

One point to remember is that there is a relation between the upper frequency limit of the amplifier and the gain setting. Using common operational amplifiers, such as the 741 used in the prototype, the upper frequency limit is about 1 MHz at the unity-gain point. It varies with frequency as shown in Fig. 7-3. Because of this gain variation, the upper frequency limit with the amplifier set for unity gain will be about 1 MHz, and with it set for +40 dB gain it will be only about 10 kHz. There will be a small increase in the upper frequency limit for minus gain

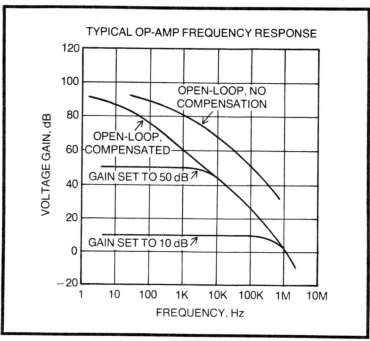

Fig. 7-3. Frequency response curve of op-amp at various gain settings.

settings, but this is small since the operational amplifier rolloff becomes more rapid above the zero gain point. If the upper frequency limit is important, it can be extended by a factor of 5 to 15 times by use of higher-performance operational amplifiers such as the LM110, the CA3130, etc.

To further increase the usefulness of the amplifier, two additional input points can be provided. One is a connection to the noninverting input of the amplifier, giving the connection of Fig. 7-2B. The gain of this connection is given by

$$\frac{e_2}{e_1} = \frac{R2 + R1}{R1}$$

This is nearly equal to the gain of the inverting configuration at the higher values of gain, when R1 is small compared to R2. However, when R1 is equal to R2 the amplifier gain is 2:1, or 6 dB. The gain cannot be reduced below unity, as it can in the inverting configuration. One useful point to remember is that the resistance values which divide the inverting gain into steps of 2 dB divide the noninverting gain into even smaller steps, with the range from 0 to +6 dB being divided into 20 steps—*but these steps are not linear.* Another point to remember when using the noninverting configuration is that the inverting input must be grounded. If it is not, the noninverting gain becomes 0 dB, since R1 becomes infinity.

The second input connection to be provided is the junction of R2 and R1, which is also the direct inverting input to the integrated circuit. Providing this point allows connection of the circuit as a summing amplifier, Fig. 7-2C. The gain relation for this connection is

$$e_2 = R2 \left(\frac{e_A}{R_A} + \frac{e_B}{R_B} + \frac{e_C}{R_C} \right)$$

For minimum error in summing, all resistors should be identical: this is the 0 dB gain condition. However, the error is usually acceptable for other values of resistance, as required for gains up to 40 dB.

The availability of the inverting and noninverting inputs allows the amplifier to be used as a differential amplifier (Fig. 7-2D). The amplifier is now sensitive to the difference between the two inputs and relatively insensitive to their individual magnitudes, the gain relation being

$$e_2 = \frac{(R_A + R2)}{(R_B + R_C)} \frac{R_C}{R_A} e_B + \frac{R2}{R_A} e_A$$

The condition for minimum error is that the parallel resistances of R_A and R2 should be equal to the parallel resistances of R_B and R_C. To simplify obtaining this condition all resistances can be chosen to be equal, which also has the desirable feature of giving the least offset error for either the inverting or noninverting connections.

Still another amplifier feature which is often needed is capacitive coupling. In the prototype, this is provided for the inverting and noninverting inputs, and for the output. For this connection the low frequency response of the amplifier will vary with gain in a different fashion than normal, since the impedance of the input capacitors enters into the gain relations given above. Approximately, for the values of the prototype, the low-frequency half-power point is 100 Hz at unity gain. The direct-coupled inverting input terminal must be grounded when the noninverting input is used, even though the noninverting input is to be AC-coupled.

CIRCUITS AND COMPONENTS

The complete diagram of the amplifier is shown in Fig. 7-4. This follows directly from the block diagrams. R1 is switchable, ten series resistors being used to give the desired value for each switch point. Separately, R2 and R3 are varied simultaneously by a two-gang switch.

The values of these resistors were chosen from the standard 5% series. Many are the same as 10% values, which could be used if exact values of gain are not important. Many

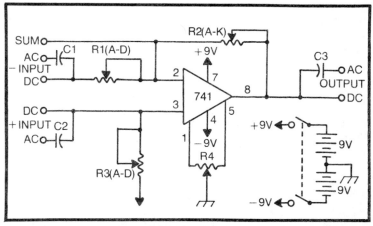

Fig. 7-4. Circuit diagram of the step-gain amplifier. R1, R2, and R3 are switch-selected to vary gain in steps. While a 741 Op-Amp was used in the prototype, many other types are usable.

Table 7-1. Gain-Setting Resistor Values.

GAIN, dB	IDEAL VALUE	PRACTICAL VALUE (5π)
dB		
0	1000Ω	1000Ω
2	259Ω	300Ω
4	326Ω	300Ω
6	410Ω	390Ω
8	517Ω	510Ω
10	650Ω	680Ω
12	819Ω	820Ω
14	1031Ω	1000Ω
16	1298Ω	1200Ω
18	1633Ω	1600Ω
20	2057Ω	2200Ω

of the values for R1 are compromises, chosen to match the ideal values as closely as possible. The theoretical values (giving exactly 2 dB per step) and practical values (chosen from standard 5% resistors) are given in Table 7-1. If high precision of gain is important, R1 should be adjusted to the ideal values either by selection from an assortment of resistors or by parallel or series connection of two resistors.

It is possible to modify the design to provide a different change in gain per step. For example, 1 dB per step would be possible, or 3 dB, or even 6 dB, the last corresponding to a 2:1 voltage change. The main reason for using 2 dB per step in the prototype was that this allowed use of standard switches and standard resistance values while providing reasonable resolution.

There are no critical areas in the amplifier. Partly this is due to the use of the 741 operational amplifier, which is internally compensated. Because of the relatively low gain, the lead layout is not critical, and it is very unlikely that oscillation problems will be encountered. If other operational amplifier IC devices are used, there may be need for external compensation resistors and capacitors. (Connection points for the most common compensating arrangements are provided by the layout shown.)

CONSTRUCTION

All components except the terminals and the power switch are mounted on a PC board. Board layout is shown in Fig. 7-5; parts location is reflected in Fig. 7-6.

Resistor switching is arranged to allow gain to increase for either clockwise (CW) or counterclockwise (CCW) switch rotation; CCW was used in the prototype. Dial markings, of

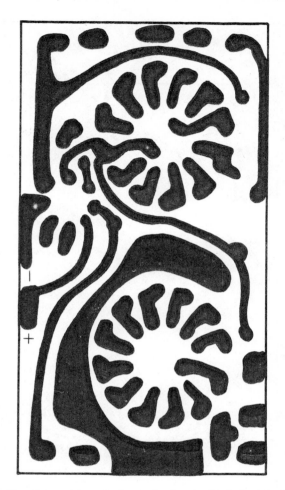

Fig. 7-5. Circuit board layout, drawn freehand with India ink. (Original board drawn freehand with an ink-resist pen.)

course, must be adjusted according to the rotational sense used. The resistor placement shown in Fig. 7-6 matches the dial shown in Fig. 7-1. The operational amplifier connections allow use of nearly all TO-5 case amplifiers without crossing leads. Spare pads are provided for mounting an offset-centering potentiometer (R4). This is not needed for most AC work, or if the bias at the DC terminal is not important.

Panel layout is shown in Fig. 7-7. The convention of minus inputs on the left, plus inputs at the top left, and output at the top right is used. The summing node is at the upper left corner. Only one ground terminal is provided—the amplifier gain is

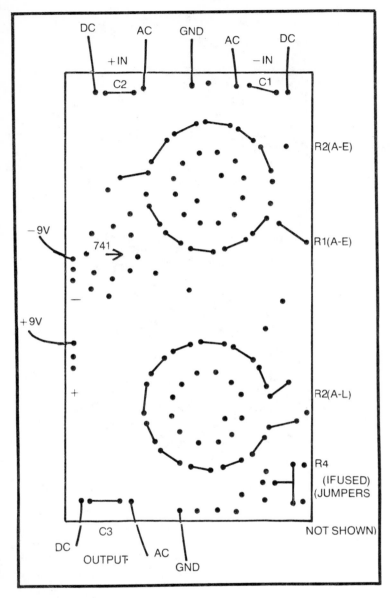

Fig. 7-6. Drilling template, component layout, and lead connections.

sufficiently low that there should be no stability problem due to circulating ground currents. In the prototype the binding posts are color coded—black for ground, red for AC-coupled connections, blue for DC-coupled, and white for the summing

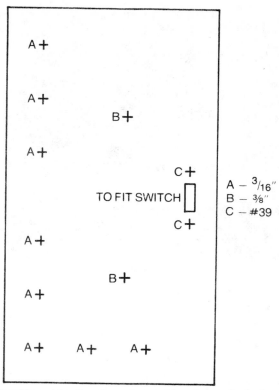

Fig. 7-7. Panel layout for the step-gain amplifier. Labeling is evident from Fig. 7-1.

node. The white and blue binding posts were made from standard insulated 5-way posts by painting the plastic with spray paint.

Dial markings are selected to give unity gain with the two indicator dials at zero and with $R1 = R2 = R3$. With the resistor arrangement shown, this requires minus markings on the *fine*-switch dial (left dial on panel photo of Fig. 7-1). The gain is thus the algebraic sum of the dial settings.

As with other instruments of the miniature laboratory series, this unit is intended for external battery power supply. Pigtail leads are provided for the two 9V supplies needed. There is ample room within the case for two 9V batteries or for a small AC supply pack. Note that the amplifier will operate properly over the range from $\pm 6V$ to $\pm 18V$, with essentially no change in amplifier gain. There will, of course, be a change in maximum signal handling capability, which is limited to approximately 80% of the total supply voltage.

Fig. 7-8. Internal view, showing demounted PC board. There is ample room in the case to mount the two 9V batteries required for operation.

Other details of construction can be seen from Fig. 7-8, which shows the circuit board removed from the chassis, an aluminum minibox of 2 × 3 × 5 inches.

ADJUSTMENT AND CALIBRATION

There are two options open for adjustment and calibration of this amplifier. One is to use resistors as specified. If this is done, the gain can be expected to be within 10% of the switch-indicated value and probably to be within 5% of the value. If it is necessary or desirable to know the gain more accurately, the values of the individual resistors can be measured, substituted in Table 7-1, and the gain for each setting of the switch calculated.

An alternative is to measure the gain after assembly. This can be done using the precision potentiometer of the RC(L) bridge as a voltage divider, matching the amplitudes on an oscilloscope or meter, and reading the gain from the precision potentiometer. The other basic alternative is to select resistance values to correspond to the theoretical values of Table 7-1, either by selection from an assortment of resistors, or by using series and parallel connections to get the exact value. Using the RC(L) bridge for this should give a gain error of better than ±1%.

For the prototype, the resistors were selected from a lot of 5 and 10% resistors, the nearest to the 5% value being used. The measured values were recorded and the gain calculated.

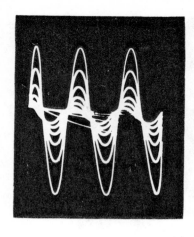

Fig. 7-9. Multiple exposure of output waveform, showing amplitude differences for several gain settings (input constant).

USE OF THE AMPLIFIER

One use of the amplifier is for general-purpose amplification. Its wide range of gain plus the fact that signals can be maintained at the same polarity or inverted in sign makes it a useful general-purpose amplifier. It is especially useful for the DC and audio range when used as an oscilloscope preamplifier. Figure 7-9, a multiple exposure at various gain settings, symbolizes the ease of use.

A second use of the amplifier is as a precision attenuator, capable of supplying gain or loss. For example, a 1.55V calibrating dry cell provides a 15.5 mV calibration signal at maximum attenuation. Other voltages, up to a maximum of about 6V with ±9V supply, are also available. The AC range can be used in the same way. Both the AC and DC connections are useful with a low-cost oscilloscope that does not have a precision attenuator.

A third use for the step-gain amplifier is in the summing connection. It is especially useful for waveform generation, by using the basic waveshapes from the function generator of Chapter 9. For example, the square-wave output of the generator can be added to the triangular output to form the trapezoid required to drive magnetically deflected oscilloscopes. Two equal-amplitude sine waves can be added together to test the intermodulation distortion of an audio amplifier, or the peak output of an RF or audio amplifier. Figure 7-10 shows just one of many possible waveforms, created by summing a sine wave and negative square wave of the same frequency.

A fourth use of the step-gain amplifier is as a differential amplifier. An example of this use is in the study of distortion in

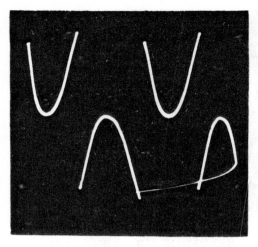

Fig. 7-10. Example of summing of two signals, one a 60 Hz sine wave and the other a 60 Hz square wave (peak amplitudes equal).

amplifiers. The output contains the input wave plus the distortion generated within the amplifier. By feeding a sample of the output to one of the differential inputs, and a sample of the input to the other, and adjusting the sample level to cancel the input wave, the distortion can be displayed on an oscilloscope. The same technique is useful in DC measurements, where a change of a few millivolts DC is biased by several volts, or where it is obscured by the simultaneous presence of a large AC voltage.

A major possible use is in the field of analog computation. Many of the computation modules needed in a typical problem are immediately available in the step-gain amplifier. Many more can be provided with a few external components, a potentiometer for continuously variable coefficients, or a capacitor for integration, for example. The function generator of Chapter 9, a half dozen step-gain amplifiers, and the lab oscilloscope will give a good beginning to full analog computing capability.

STEP-GAIN AMPLIFIER SPECIFICATIONS

- Gain adjustable: −40 to ×40 dB, in 2 dB steps
- Frequency range: 0−10 kHz (maximum gain)
  0−1 MHz (unity gain)
- Inverting or noninverting polarity.
- Usable as: integrator, differential amplifier, summing amplifier

PARTS LIST FOR STEP-GAIN AMPLIFIER

C1,2,3—0.1 μF Mylar capacitor, 100V
R1, R3:   All resistors ½W
A — 1K
B — 10K
C — 100K
D — 1M

R2:
A — 1000Ω
B — 300Ω
C — 300Ω
D — 390Ω
E — 510Ω
F — 680Ω
G — 820Ω
H — 1000Ω
I — 1200Ω
J — 1600Ω
K — 2200Ω
R4 — 10K potentiometer
IC — μA741, TO-5

Chapter 8

The Tone-Burst Generator

An instrument not commonly found in small laboratories is the tone-burst generator; but every experimenter, on occasion, needs the ability to interrupt a wave train at regular intervals, to turn it on and off repetitively. One common need is to reach the peak output capacity of an amplifier without exceeding its average rating, a problem in testing hi-fi equipment or amateur linear amplifiers. Another common need is generation of a burst signal, radar-like in form. Still a third need is to provide a train of a known number of pulses, to simulate a digital word.

This tone-burst generator (Fig. 8-1) is intended to provide these off/on trains of signals while maintaining simplicity. Virtually any input wave can be used. The instrument can be

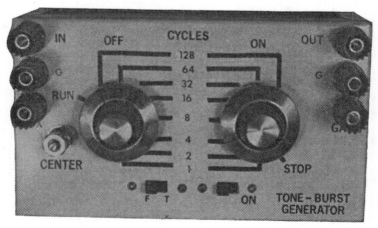

Fig. 8-1. The tone-burst generator. One dial sets the number of cycles in the burst, the second the number omitted. Panel finished in black lettering on white background.

set to produce a specified number of cycles, and then to be off for a separately specified number. Alternatively, the instrument can be set to turn a wave train on for a specified time period (milliseconds as designed), and then turn the train off for a separately specified time. In addition, a second wave train can be used to control the first, determining the off/on cycle. Auxiliary terminals are provided, one being for the conversion of the input wave into a square wave, the other being for the wave train which represents the off/on cycle.

THEORY OF OPERATION

As can be seen in Fig. 8-2, the instrument includes four basic parts. The first is a gate which serves to interrupt the input wave train or to allow it to pass to the output. The second element is a *zero-crossing detector*, which changes from plus

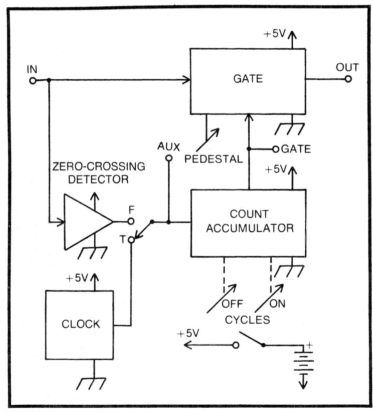

Fig. 8-2. Block Diagram of the tone-burst generator, showing major elements, interconnections, controls, and switching. Output bursts may be controlled from the input or from an internal clock.

to minus, or reverse, each time the input wave crosses its average value. The third part is a counting circuit, which accumulates the number of cycles or zero-crossing events and which signals the gate off/on action. The final element is a simple clock, which provides the alternate input to the counting circuit when time rather than cycle count is to be the controlling factor.

The gate is a specialized voltage divider, as shown in Fig. 8-3. There are two divider arms, one for the *on* cycle and the second for the *off* cycle. During the *on* cycle the magnitude of one resistance (that of the FET) is controlled by the input signals, so that the output is a replica of the input. During the *off* cycle this FET arm is essentially open-circuited, and is replaced by an adjustable resistor (R6). The value of R6 can be set to give an output equal to the average value of the input waveform or it can be adjusted to place the output signal on a pedestal. The gate FET is protected against overload by a zener protective circuit.

The counter circuit (Fig. 8-4) includes three basic parts. The first is a straight digital counter of eight stages, which can

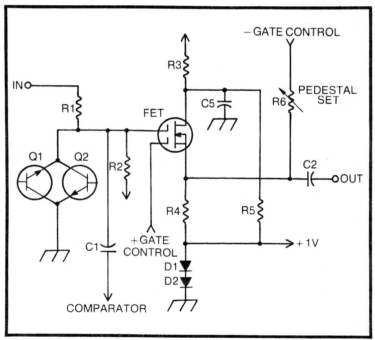

Fig. 8-3. Circuit diagram of the gate, a special form of voltage divider. Transistors Q1 and Q2 are used as protective diodes, and may be omitted if a gate-protected FET is used.

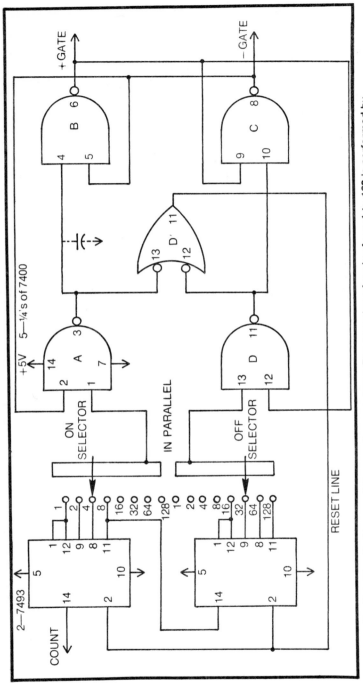

Fig. 8-4. Circuit diagram of the count accumulator. The counting of cycles from 1 to 128 is performed by the two 7493 counters for both the **on** and **off** cycles. The five TTL dual gates perform the switching and gate generation functions.

accumulate up to 128 counts. The next element is a *count-value* detector, which generates a reset signal whenever the preset number of pulses is accumulated. Finally, there is a switchover circuit that generates the *off* gate while causing the counter to accumulate *off* cycles, then generates the *on* gate while accumulating *on* cycles. The steering elements in the switchover circuit ignore the inactive cycle setting.

The zero-crossing detector (Fig. 8-5A) is an IC comparator plus a Schmitt trigger. Since the comparator input is AC-coupled, the zero-crossing reference is the average of the input waveform.

The clock, shown in Fig. 8-5B, is a simple multivibrator set to give a period of 1 msec. If desired, the clock period can be adjusted to other values (for example, to give 1 sec periods).

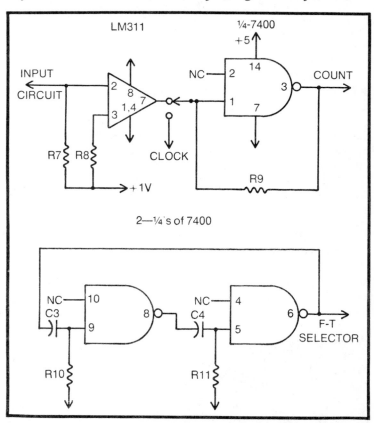

Fig. 8-5. Additional circuit elements: (A) zero-crossing detector, with Schmitt trigger as driver; (B) multivibrator clock, for timed bursts of signals.

107

The interconnection of these circuits is evident from Fig. 8-2, which also serves as the interconnection wiring diagram.

The output will be in the form of bursts of the input, where a burst may be synchronous with the input, and of a set number of cycles, or where the burst may be asynchronous, lasting for a preset time interval. The *off* period is also either synchronous or timed. The selectors provide *hold* and *run* operations, for no output or for a continuous wave train. The instrument output level is approximately 80% of the input level, with the total excursion of signal plus pedestal limited to approximately 1.0V.

The GATE terminal makes the positive control gate available externally. It is always synchronous with the burst. Another terminal provides either a squared-up version of the input or the clock output. In addition, when the F-T selector is set to its center or no-connection position, a square-wave input signal can be injected into the output terminal, to control the count accumulator action. This allows modulation of the regular input signal by the second signal, in a number of ways.

The circuits have been kept simple and no isolation or buffers are provided. In use, attention must be given to signal levels and load and source resistances.

Figures 8-6 and 8-7 show typical wave trains which can be developed. Figure 8-6 shows a sine-wave pulse train of two cycles off, two cycles on. In Fig. 8-7 the wave train is two cycles off, four cycles on, with the *on* cycles sitting on a pedestal approximately equal to the peak-to-peak amplitude of the wave train. The *off* and *on* durations can be set to give cycle counts between 2^0 and 2^7, or 1 to 128 cycles.

COMPONENTS AND CRITICAL CIRCUITS

The counting elements and associated circuits use TTL digital integrated circuits, readily available at low cost from a number of sources. There are no special problems in their use. The IC for the zero-crossing detector must be chosen from a limited number of types if satisfactory operation is to be

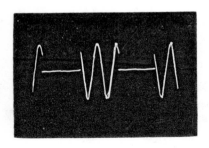

Fig. 8-6. Oscilloscope photo, two cycles on, two cycles off, sine-wave input, and no pedestal. Peak-to-peak input amplitude is approximately 600mV. input.

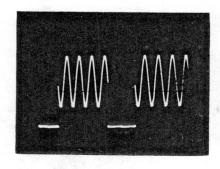

Fig. 8-7. Oscilloscope photo, four cycles on, two cycles off, sine-wave input. Input amplitude is approximately 600 mV p-p (pedestal about 500 mV).

secured with a 5V supply. Actually, any of the commonly available comparators could be used if the required additional voltages are supplied. Even the common 709 or 741 operational amplifiers could be substituted for the comparator, but with a loss in upper frequency response. This complexity of an additional supply was not desired for the prototype, however.

The gate element can be almost any type of double-gate FET. The 3N128 (available in surplus) was used in the prototype; however, with the component values shown, not all of the surplus units will be satisfactory—selection of the 3N128s or adjustment in component values may be needed. It should be noted that the 3N128 has no internal gate protection and is therefore easily damaged by static electricity. Experimenters who are in dry areas or in an air-conditioned building may want to use a device with built-in gate protection, such as Sylvania's ECG-222. The prototype has been tested with these and functions properly.

There is one critical circuit in this unit—in the set/reset multivibrator chain, which generates the *off/on* cycle. The problem occurs when both selectors are set to give the same cycle count. In this condition, a race occurs. The solution to this is to add a holding capacitor to one of the set/reset multivibrator circuits. The location and size found satisfactory for the prototype are shown on the diagram.

Since the comparator is operated without feedback, there should be no problem in oscillation with it. However, due to stray wiring capacitance, and to the wide variation in characteristics of surplus units, oscillation might be encountered, as was found in one test comparator of the prototype. If this should occur, a few picofarads of capacitance from the comparator output to its inverting input should neutralize the circuit and stop the tendency to oscillate.

All components of the comparator are available from mail-order surplus houses, and virtually all components are available from the chain radio stores.

Fig. 8-8. Internal view of the tone-burst generator. The capacitor mounted on the foil side of the board is C6.

CONSTRUCTION

Physical construction of the tone-burst generator is standard for units of this miniature lab series. As shown in Figs. 8-8 and 8-9, the PC board is supported by the selector switches, which are single-deck units with the terminals at right angles to the deck. Both Allied/Radio Shack and Lafayette carry the proper switch. Prepare these as shown in Fig. 2-9 (Chapter 2). Be careful that all terminals are clipped on the same side—which side is not important, but they must all be the same.

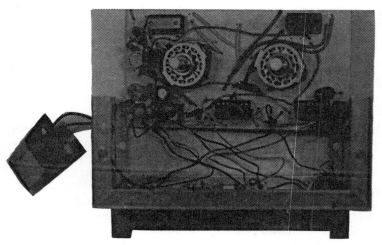

Fig. 8-9. Disassembled view. The printed circuit is supported by the two selector switches. The jumpers could be eliminated by using a double-sided printed circuit board.

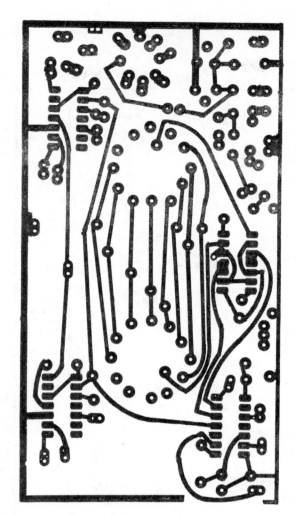

Fig. 8-10. Circuit board layout, prepared with commercial drafting aids.

The board layout used is shown in Fig. 8-10, and the component mounting in Fig. 8-11. Wiring and jumper indicators are shown in Fig. 8-12. Note that this board uses a number of jumpers. These could be eliminated if double-sided board is used, at the expense of some complexity in layout.

The TTL integrated circuits can be soldered in place, or they can be plug-in devices by using terminals. (Sockets were used for the prototype.) The TO-5 comparator is soldered in after forming leads over a half-inch-diameter rod as shown in Fig. 2-8.

The panel layout, evident from the photo of Fig. 8-1, is shown in Fig. 8-13. The centering control could be better

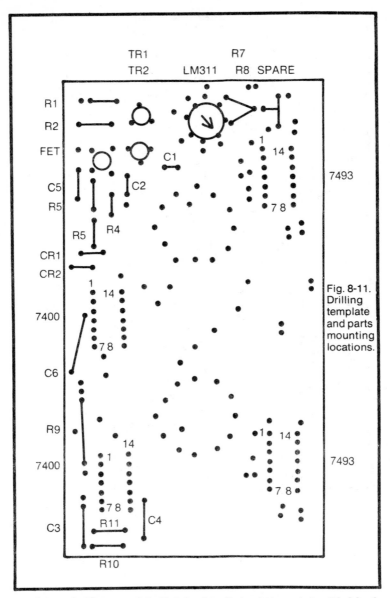

Fig. 8-11. Drilling template and parts mounting locations.

positioned. The panel is finished in flat white paint with black lettering. The black masking tape used in PC board layout is used for the heavy lines representing switch positions. When drafting is finished, the panel is sprayed with transparent lacquer to protect the lettering and tape. Binding posts are painted in three colors for easy identification.

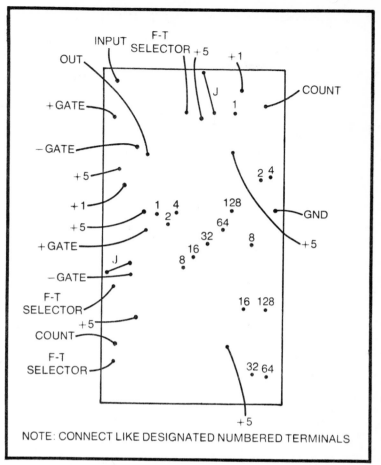

Fig. 8-12. Lead arrangement and jumper indication: the two points having the same number are to be jumpered (8 − 8, 16 − 16, etc.).

In common with the other instruments of the miniature lab, the unit is provided with pigtail leads for power connection. There is ample space within the cabinet to mount a small 6V battery. If this is used, a 1A silicon diode should be placed in series to give approximately 5.3V for operation of the ICs (each silicon diode exhibits a drop of about 700 mV). This also gives protection against application of reverse polarity.

CHECKING AND ADJUSTING

First, check for errors in construction, paying particular attention to possible shorts. Inspect all terminal pads for possible cold-solder connections. Check terminal resistances,

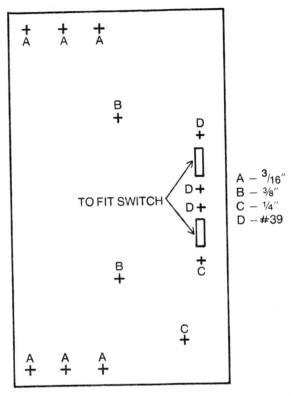

Fig. 8-13. Panel layout of the prototype. The location of the pedestal control and the spacing of the binding posts could be improved. Lettering is evident from Fig. 8-1.

then apply power and check at each +5V IC terminal and at the +1V bus.

Start the operating check with the clock (timing multivibrator). The output as seen on the scope should be a square wave, with a swing between approximately +0.3V and +4.7V, and with a period of about 1 msec. If necessary, adjust the values of C3 and C4 until this (or any other desired period) is obtained. Set the selector switch to the *time* position.

Place the selectors on the 128 position and check the propagation of the timing signal through the chain composed of the two 7493 counters. Each of the output terminals of this counting chain should show a square wave of progressively lower frequency. Check propagation of this signal through the set/reset blocks. Both a positive gate and a negative gate should be present. Check that both of these are present at the FET circuit, and that one is present on the gate output terminal. Note that if C6 has been omitted or is temporarily disconnected, the counter circuits work properly as long as the selector switches are set for a different number of *off* and *on* cycles, but that proper operation cannot be obtained with selectors on the same cycle count. For maximum count-speed capability, try reducing the value of C6, using the smallest value that gives reliable operation. Check that operation stops on the *hold* position, and is continous on *run*.

Now apply a sine-wave input signal, about 500 mV at 60 Hz, and set the selector to frequency. Check the output of the comparator. It should also be a square wave swinging from approximately 0.5V to 2.5V. If there are signs of oscillation in the comparator, place a small capacitance from its output to its inverting input. In severe cases of oscillation it may be necessary to use another comparator.

Check the FET output. Surplus FETs may show excessive leakage, insufficient output, or improper gating, and it may be necessary to try more than one unit. With the centering control set so that the *on* output and the average of the signal are at the same level, the output should have the same waveshape as the input, at about 80% of the input amplitude. The leakage during the *off* cycle should be no more than a few percent of the input voltage. Adjustment of the centering control should permit placing the *on* cycle on a positive or a negative pedestal of approximately 500 mV amplitude, without affecting the signal waveform. Too large a pedestal will affect the leakage or the waveform.

Check the output signal carefully. It should be possible to display one complete cycle of the input, or to remove one cycle, or to generate bursts or gaps of 1 to 128 cycle duration. At low frequencies each burst should start at the zero crossing of the sine wave, as in Fig. 8-6. At high frequencies, above about 20 kHz, there may be a phase shift, causing a delay as shown in Fig. 8-14. If this is important, a phase control can be set up by returning the noninverting input of the comparator to a potentiometer connected between ±1V and ground. This could be placed on the front panel, if desired.

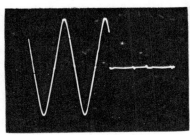

Fig. 8-14. High-frequency burst, 2 cycles on, 4 cycles off. Due to phase shift, switching occurs at about 60° after the zero crossing. A phase control can be added to eliminate this shift.

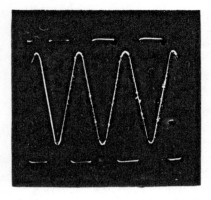

Fig. 8-15. Input and output of comparator circuit, a dual exposure with common baseline. This squaring circuit operates into the RF range.

Check that the output and the gate signal are in phase, as shown in Fig. 8-15. Check also that the signal at the auxiliary output terminal is a squared-up version of the input, as shown in Fig. 8-16. (The auxiliary output will be the timing signal when the selector switch is set to time.)

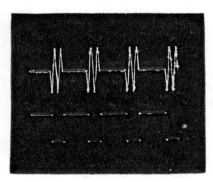

Fig. 8-16. Output signal and gate output. A dual exposure with shifted baseline. The envelope of the output and the gate will be in phase.

The gating action should be reliable. For example, Fig. 8-17 was made with an input signal of 1000 Hz, modulated 95% with a 2 Hz sine wave. Even with this wide excursion of input amplitudes, the count of 8 off, 4 on is maintained.

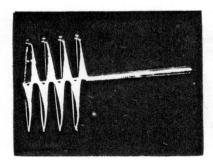

Fig. 8-17. Output of tone-burst generator for input 500 Hz sine wave modulated 95% by 10 Hz sine wave. Burst setting 4 cycles on, 8 cycles off. Scope synchronized by gate signal. There is a small shift in phase as input signal varies in amplitude, but the output is error-free.

CALIBRATION AND STANDARDIZATION

The only two values which need to be calibrated in this instrument are its transfer gain and the length of the timing cycle. Check the transfer gain using a peak-reading scope or voltmeter. This should be approximately 0.8. For the timing clock it is convenient to determine the total period, set to 1.0 msec for the prototype, plus the length of the *off* and *on* cycles of this period. These were approximately 0.45 and 0.55 sec in the prototype. The length of these could be adjusted to precisely 500 msec if desired. It may be useful to record the DC levels at the output terminals.

USES OF THE TONE-BURST GENERATOR

Some of the major uses of the tone-burst generator are mentioned in the introduction. Probably the most common use is in peak output testing of amplifiers, where a signal with a high *peak* and low *average* is needed. A typical test for music would be 8 cycles on and 64 cycles off, to give a duty cycle of 12%. In transmitter testing, the peak-to-average ratio for speech is often taken to be 13 dB, a duty cycle of 5%. This ratio can be simulated with several settings of the selectors—for example, with four cycles on and sixty-four cycles off.

Workers in sonar and other users of radar-like signals may find 4−8 cycles on and 128 cycles off useful. For this application, it may be desirable to make one of two changes in the design: One modification would be to change the time constants of the timing multivibrator, to give a very short *on* period and a very long *off* period, better for simulating radar or sonar duty cycles. An alternative to this would be to modify the input to the counting chain by adding a pair of two-input AND gates. One of these would gate from the input signal to initiate the *on* cycles, and the other would gate from the clock to initiate the *off* cycle.

One specialized use of the tone-burst generator is *harmonic generation*. The relative strength of harmonics is determined by the selector switch settings, in accordance with the equations in Table 8-1. As an example, if the switch is set for one cycle on and four cycles off, the fifth harmonic of the output will be approximately twice as strong as the fundamental. The tenth harmonic will be missing. The relative harmonic strength varies with the phase of the signal at the time of switching.

Another specialized use of the tone-burst generator is as a digital, or switching, modulator. For this, the lower-frequency signal should be placed on the regular input terminal, and the

Table 8-1. Tone-Burst Frequency Content.

For a sine wave signal, of N full cycles on duration and M full cycles off duration, the output is

$$e_s = e_{in} \sum_{n=1}^{\infty} |a_n \sin \left[\frac{2 \, nt}{(N + M)t} \right]$$

Where: e_s is output signal
n is harmonic number
a_n is given by

$$a_n = \frac{\cdot N}{N + M} \left[\frac{\sin X}{X} \mp \frac{\sin Y}{Y} \right]$$

$$X = 2N \left(\frac{n}{N + M} - 1 \right) \frac{\pi}{2}$$

$$Y = 2N \left(\frac{n}{N + M} + 1 \right) \frac{\pi}{2}$$

e_{in} = input signal level

As an approximation, the harmonic with the greatest amplitude is given by N + M: for example, for 8 cycles on and 8 cycles off, the 16th harmonic is the largest; the 15th and 17th are within 3 dB of its value.

higher frequency signal on the auxiliary terminal. The frequency—time (F-T) selector should be at the middle, or neutral position. For many applications it will be necessary to filter the output signal to remove the modulating component. The filter may be a bandpass type if sine-wave output is wanted, or a high-pass type if square wave output is needed. There are numerous possibilities of modulation by noise or noise-like signals—for production of a random-period signal, as an example.

Some additional uses are the excitation of auditoriums to determine the decay time, and thereby the need for acoustic treatment; determination of the transient response of such components and circuits as filters, meters, automatic gain control stages, etc.; the synthesis of time ticks, by initiation of a tone burst at a specific time; and measurement of human response to various transient or interrupted signals.

The tone-burst generator is far more useful than first appears; it is likely that you will wonder why you did not have one before.

TONE-BURST GENERATOR SPECIFICATIONS

- Converts continuous-wave signal into pulses: m cycles on, $m = 1-128$ binary; n cycles off, $n = 1-128$ binary

- Provides timed pulse train: x msec on, $x = 1-128$ msec; y msec off, $y = 1-128$ msec
- Sets pulse on positive, zero, or negative pedestal
- Provides synchronous gate
- Serves as zero-crossing detector
- Serves as digital modulator
- Serves as harmonic generator

PARTS LIST FOR TONE-BURST GENERATOR

C1 —	0.01 μF ceramic disc capacitor
C2 —	0.1 μF Mylar capacitor
C3 —	0.005 μF ceramic disc capacitor
C4 —	0.005 μF ceramic disc capacitor
C5 —	10 μF, 15V electrolytic
C6 —	See text
R1 —	10K All resistors ½W
R2 —	100K
R3 —	100Ω
R4 —	1000Ω
R5 —	15K
R6 —	10K potentiometer
R7 —	100K
R8 —	100K
R9 —	3300Ω
R10—	3300Ω
R11—	3300Ω

ICs:
2 — 7493 binary counters
2 — 7400 quad 2-input gates
1 — LM311 comparator

FET — 2N4220, etc.
Q1, 2 — 2N3393, etc.
D1, 2 — silicon, 1N914 etc.

Chapter 9

The Function Generator

Over the years the name *function generator* has come to mean a special signal generator whose output varies in special ways as a function of time, in some specified fashion. Usually several such functions, or waveforms, are provided. The name *function generator* came from analog computer usage, where these specialized waveshapes are used as inputs or excitations to the analog computing element. The function generator could perfectly well be called a waveform generator.

BASIC PURPOSE

The simplified function generator shown in Fig. 9-1 is intended to provide two basic waveforms plus several variations. The fundamental waveshapes are the square and

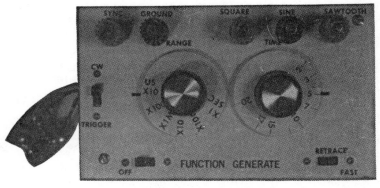

Fig. 9-1. The function generator gives square, sawtooth, and sine waves, plus variations. Calibration is for period, from 20 μsec to 20 sec. Finished in black lettering on white background. Dials are homemade, as described in Chapter 2.

the triangular wave, which are locked together in phase. Also provided is a modification of the triangular wave to give a close approximation of a sine wave. In addition, a switching circuit is included to change the slope of the triangular wave when it is going in the positive direction, to give a close approximation to a sawtooth wave; at the same time, the square wave is converted to a pulse. The sawtooth is also modified to give a sine wave, but instead of being continuous $(0-360°)$ the sine wave is interrupted and covers the half-cycle from $-90°$ to $+90°$ repetitively. There are provisions for synchronizing the signal to an external source and for triggered operation (in which each new cycle must be triggered by an external signal). This last feature allows the function generator to generate oscilloscope sweep signals, and to provide blanking to the scope at the same time. The specific waveforms and their characteristics are shown in Table 9-1.

THEORY OF OPERATION

The heart of the function generator is an op-amp integrator. The basic circuit of such an integrator is shown in Fig. 9-2A. Because of feedback through the capacitor, the output/input relation of this circuit is:

$$e_{OUT} = \frac{1}{RC} \int_{t_1}^{t_2} e_{IN} \, dt$$

in other words, integration of the input signal.

To secure the desired triangular output, the input of this integrator must be fed with a square wave. This is readily generated by a special form of trigger circuit, sometimes called a bistable flip-flop or *threshold detector*. The circuit, shown in Fig. 9-2B, is based on use of an op-amp as the active element. The circuit has feedback from the output to the

Table 9-1. Function Generator Waveforms.

WAVEFORM	CHARACTERISTIC	AMPLITUDE	MODE[1]	FULL PERIOD
square wave	symmetrical	15V p-p	N	4 μsec to 40 sec
triangle wave	symmetrical	10V p-p	N	Same
sine wave	$0-360°$	2V p-p	N	Same
sawtooth	1% return[2]	15V p-p	F	2 μsec to 20 sec
pulse	1% duty cycle	10B p-p	F	Same
sine wave	$-90°$ to $+90°$	2V p-p	F	Same

[1] N = normal; F = fast response
[2] 1% of maximum period of range setting

All wave forms can be triggered, or synchronized to an external source.

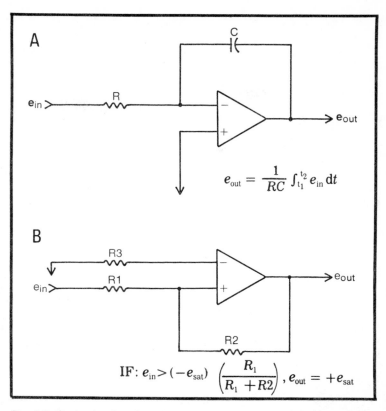

Fig. 9-2. Basic circuits of the function generator. (A) Op-amp integrator used to generate triangular waves by integrating a square wave input; (B) two-state amplifier or threshold detector, used to generate square-wave output whenever triangular wave crosses zero reference.

noninverting input of the operational amplifier—in other words, this is a positive-feedback circuit. Also, there is a summing circuit connected to this input, which adds together the feedback voltage plus an external voltage. Whenever the external voltage component is of opposite sign to the feedback voltage, and larger in magnitude, the output changes sign. The speed of the change is determined by the internal characteristics of the op-amp. The op-amp output then holds this new sign of output until the input signal again causes the voltage at the op-amp noninverting terminal to change sign.

The desired square-wave switching action is secured by connecting these two circuits in a feedback loop, as shown schematically in Fig. 9-3A. The threshold detector now changes sign whenever the integrator output exceeds a certain magnitude of signal. Since the integrator action is

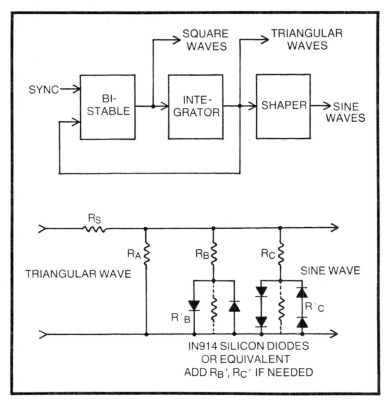

Fig. 9-3. Details of the function generator: (A) feedback connection of the basic circuits of Fig. 9-2—circuit is self-starting; (B) arrangement of shaper circuit shown in part A. (Diode switches connect added elements into a voltage divider and serve as nonlinear resistances to convert a triangular wave into a sine wave.)

symmetrical, the triangular-wave output from it will also be symmetrical and the threshold-detector output will follow this symmetrically, forming a symmetrical square wave, the required input to the integrator.

The switching points are set by the ratio of R2 and R1. This determines the amplitude of the triangular wave. The frequency is separately controllable by varying the value of the quantity RC in the integrator. In this function generator, C is varied by a factor of 10 to give decade steps of frequency, and R is continuously varied (also over a factor of 10) to give interpolation between the decade steps. (If desired, half-decade steps could be used instead, by doubling the number of capacitor values and changing the value of R to give $\sqrt{10:1}$ variation in total resistance.)

To extend the usefulness of the function generator, some additional circuitry is provided. The first addition allows for synchronization and for triggered operation. In Fig. 9-2B, the reference point for the threshold detector is the zero-voltage point established by the ground end of resistor R3. This resistor could be returned to some other voltage which is larger than the switching point set by resistors R1 and R2. In this case, switching will not occur and operation of the circuit will cease. Operation can be initiated by feeding a short pulse of proper polarity to the inverting input. This will cause the threshold detector to change state, which then causes the integrator to start another cycle of operation. This cycle will continue to completion, but another cycle will not start automatically, since it is again stopped by the fact that its reference point is beyond the range of the integrator. Action stops again, and remains stopped until another sync pulse occurs. The only restriction is that the sync pulse length be short compared to the total length of the cycle.

Having this trigger input makes possible another action: The threshold detector changes state whenever the voltage fed to it exceeds the reference voltage. The sync input allows change of this reference point by an external voltage, and so allows synchronization of the operating cycle to an external source. As is common with such sync circuits, the period set internally must be a little longer than that of the external signal if proper action is to be obtained. If this relation does not hold, the circuit will attempt to sync every cycle, thus giving a distorted waveform.

To allow adjustment of sync sensitivity, the reference voltage at the end of R3 can be made adjustable. The potentiometer for this can also be used to set the average value of the triangular wave. It can be used, for example, to give an output triangle with an excursion of $+1V$ to $-1V$, or an excursion of $+3V$ to $+1V$, or some other value. This potentiometer could be placed on the front panel if desired, although in the prototype it is an internal adjustment and used only for setting the trigger threshold.

In addition to triangular- and square-wave functions, sine waves are often needed. Triangular and sine waves are close relatives: A sine wave starts at the same slope as a triangular wave, and departs very little from the triangular for about an eighth of a period. Then the sine wave starts to lag behind the triangle and by one-fourth period is only 70% of the amplitude of the triangle.

It would be possible to form a sine wave from the triangular by providing a continuously adjustable potentiometer which is on its full-scale point when the triangle wave is at zero voltage, and which decreases to 70% of full scale when the triangle is at its peak. The action of this potentiometer can be approximated by switching in resistors as the voltage of the triangular wave increases. Figure 9-3B shows such a switchable potentiometer, very nearly the simplest possible of such types. It uses the internal voltage drop of diodes to set the switching points, and the diodes themselves as the switches. In principle, only three slopes are possible, but because of the nonlinearity of diodes, very close approximation to a sine wave is possible by selecting proper values of resistance.

There are many applications in which it is desirable to have a signal which has a certain slope to a maximum point and which then returns rapidly to the starting value—an example is an oscilloscope sweep. This can be accomplished in the function generator by providing one set of integrator RC time constants for the positive part of the cycle and another set of RC time constants for the negative part. Switching from one to the other can be accomplished with diodes. The second time constant could be set at a specific value or could be made variable. For simplicity, the prototype uses a fixed value, of about 1% of the maximum period of each 10:1 period setting.

CIRCUIT AND CRITICAL ELEMENTS

The complete circuit assembling these various functions is shown in Fig. 9-4. In the prototype, six decade ranges are provided for the integrator, as established by switched capacitors. Within each decade the period is continuously adjustable, as set by the series combination of R6 and R7. The only other adjustment is the threshold setting for triggerable operation—an internal adjustment. *Normal* or *fast* retrace is panel selected, D1 and D2 being in parallel for normal operation, with R8 not connected.

The prototype makes no provision for amplitude adjustment (which would require R9 to be variable), nor for return time-constant adjustment (which would require R8 to be variable). The threshold setting adjustment (R3) could be placed on the front panel, where it could serve as a centering adjustment as well as for setting the trigger level.

The controls which are provided are labeled RANGE for the selection of decade capacitors, and PERIOD for the adjustable resistance. (In the prototype this is calibrated for half-period; in other words, for the duration of one positive-going cycle. Since the wave is symmetrical for normal operation, this is also the duration of the negative going cycle.) The switch

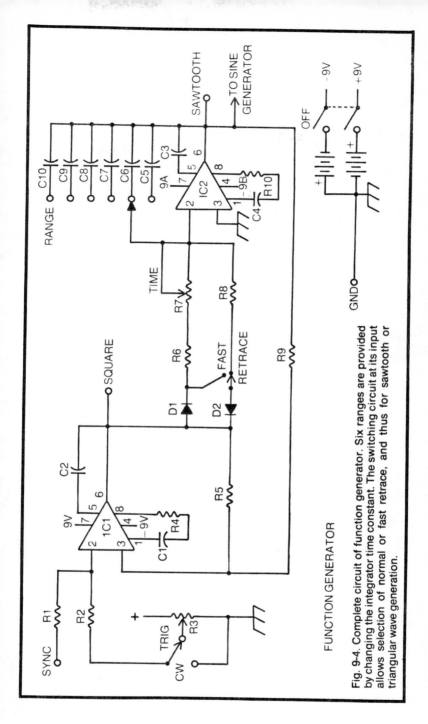

FUNCTION GENERATOR

Fig. 9-4. Complete circuit of function generator. Six ranges are provided by changing the integrator time constant. The switching circuit at its input allows selection of normal or fast retrace, and thus for sawtooth or triangular wave generation.

provided for continuous or trigger operation is labeled CW/TRIGGER, and that for *fast* or *normal* retrace is labeled RETRACE.

Since the square-wave amplitude is established by saturation of IC1, its level will change with battery voltage. This will affect the period. See the first reference for corrective action, if this is important.

None of the components in the function generator are especially critical. Potentiometer R7 should be chosen for low backlash and low noise. The specified type of control satisfies these requirements, but many wirewound or carbon variable potentiometers are satisfactory. For the *range* capacitors, the best type would be polystyrene for the small values, Mylar for the intermediate ones, and tantalum for the large value. Satisfactory operation can be obtained with mica or ceramic units for the small values, and good-quality paper for intermediate values. Aluminum electrolytics can serve for the large values, but there may be some problem with leakage unless the quality is very good.

The upper frequency limit is set by the op-amps used and is approximately 20 kHz with the commonly available 709 and 741 devices. It may be extended to approximately 100 kHz by use of such types as the LM301, and to approximately 300 kHz by RCA's more recent CA3130.

CONSTRUCTION

The construction of the function generator follows the standard for the miniature laboratory series. All components except terminals and three switches are mounted on a PC board. The board is mounted within a small aluminum chassis-box of about $2 \times 3 \times 5$ inches. In this design the board is supported within the case by posts.

Board layout is shown in Fig. 9-5. The parts placement for this is shown in Fig. 9-6. Note that the terminal pads for R7 have been made oversize; these should be placed to fit the control used. The ones shown are centered for an Allen-Bradley potentiometer mounted through the board, and connected to the pads with wire leads. These pads will also fit most PC-type variable resistors.

The board layout for the printed circuits is intended to allow choice of amplifier type. In the prototype, 709s were used, with compensation values as shown; 741s would be satisfactory, and would allow elimination of the compensation components. Other op-amp types can be used, some requiring different arrangements of the jumpers for correct connections.

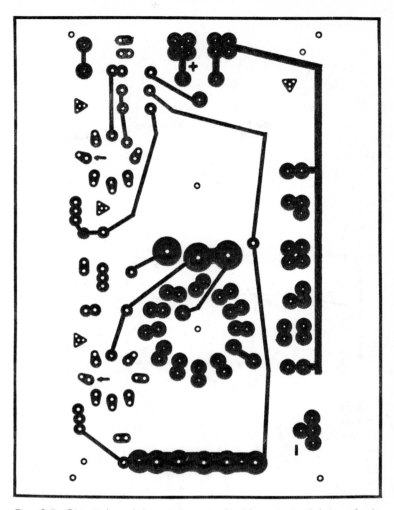

Fig. 9-5. Circuit board layout (prepared with commercial transfers). Original board drawn freehand with ink-resist pen.

When mounting the diodes, note that absolute polarity is not important unless a particular sense of the sawtooth output is desired: if it is, it is controlled by the polarity of D1 and D2. The important point is to make sure the relative polarity of the diode pairs is opposing.

For good waveform symmetry, all diode pairs should be matched for conduction and back resistance. The best procedure for selection is to first match for forward conduction, using the ×1 scale of an ohmmeter which uses a 1.5V battery. The matched units selected should then be

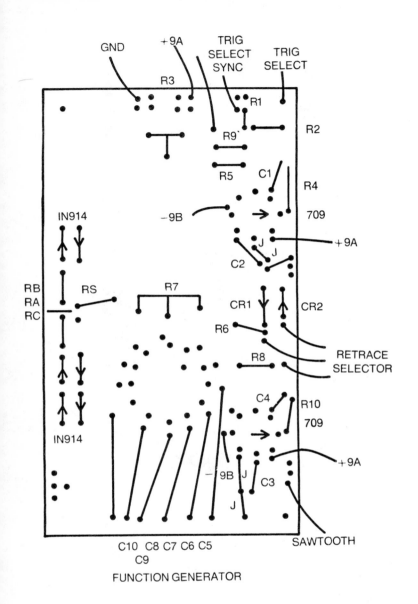

Fig. 9-6. Circuit board drilling template, parts placement, and lead connections. The jumpering for the integrated circuits allows any TO-5 device to be used.

further matched for back resistance using a high-resistance scale, preferably one that uses a 9V to 15V battery. Diodes D1

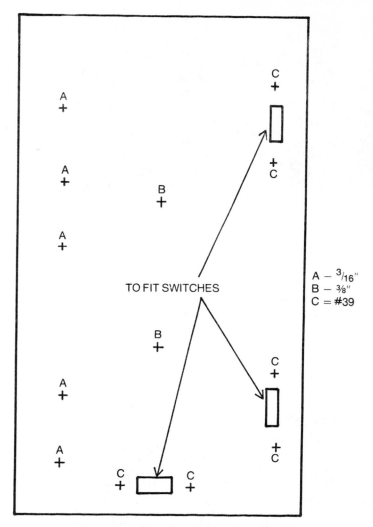

Fig. 9-7. Chassis drilling template.

and D2 should have as high a back resistance as possible—several megohms ideally; in any event, the back resistance should be at least 10 times the value of R6 and R7 in series.

If good accuracy on all scales is desired, range-setting capacitors C5—C10 should be selected to have precisely a 10:1 *ratio* of values—the *exact* value is not especially important. Probably the easiest way of obtaining 10-to-1 ratios is to select the largest capacitor, measure its value, and use this value

Fig. 9-8. Internal view, showing PC board. There is room in the case for mounting two 9V batteries.

divided by 10, 100, etc. for the smaller capacitors. You can use two units in parallel to obtain the desired value.

Panel layout is shown in Fig. 9-7 and is also evident from Fig. 9-1. The chassis preparation procedure consists of drilling, cleaning with steel wool, painting with matte white, lettering, and protecting by a coat of transparent lacquer. Red and black binding posts were used for the prototype (black for ground and red for other functions). If desired, each terminal

Fig. 9-9. Internal view, disassembled. The panel support posts are made from toothbrush handles. All components except terminal and switches mount on the board.

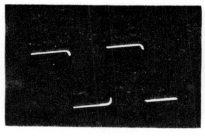

Fig. 9-10. Square-wave output. Rise time is about 5 μsec with op-amp used.

can be color-coded by spraying the binding post insulators with the appropriate color.

The internal arrangement is evident from Figs. 9-8 and 9-9. In common with most units in the miniature laboratory series, pigtail leads are used for external 9V batteries. There is ample space in the case for internal mounting of the batteries or for a small AC supply.

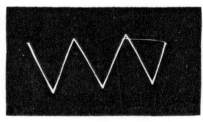

Fig. 9-11. Sawtooth output. Linearity and symmetry are better than ±1%.

CHECKS AND ADJUSTMENT

Initial work on the amplifier should be done with the operational amplifiers compensated per the manufacturer's recommendation. This eliminates any possibility of confusion due to op-amp oscillations. For initial test, set the unit for normal retrace and continuous-wave operation. If no errors have been made, the unit should start functioning immediately on application of power. If there is no output, it is likely that D1 and D2 are poled in the same direction or that one of the op-amps is incorrectly installed. A voltage check should quickly resolve the difficulty. When working properly, the

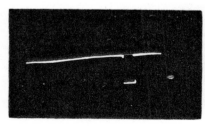

Fig. 9-12. Pulse available at square-wave terminal with fast retrace. Duration controlled by value of R8 plus forward resistance of DZ.

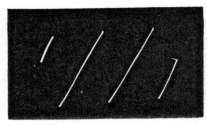

Fig. 9-13. Triangular wave, available at sawtooth terminal with fast retrace. Since periods as long as 20 sec are available, this is useful for very slow oscilloscope sweeps.

waveshapes for the square-wave and sawtooth outputs should be that of Fig. 9-10 and Fig. 9-11 for all positions of the RANGE switch and the period-varying resistor. When these patterns are obtained, set the switch for fast retrace. The waveshapes should now change to those of Fig. 9-12 for the square wave and Fig. 9-13 for the sawtooth. The square wave is now a pulse of about 1% duty cycle, and the triangular wave a sawtooth, also with about 1% return cycle. Failure to obtain these waveshapes is probably due to a defective D1 or D2 diode.

When the waveforms have been obtained, return to normal retrace. Feed a small sync voltage into the SYNC terminal. Locking of the signal to the sync should be observed, with action essentially the same as that of a normal oscilloscope sync circuit. With this obtained, remove the sync signal and set the switch for trigger operation. Adjust R3 until there is no oscillator output. Reapply the synchronizing signal and vary its amplitude. As the signal becomes larger than the bias on the op-amp, triggering should begin and then become reliable, just as in a triggered oscilloscope circuit. Triggering should be rock-steady with triggering signals having a short rise time, but may show jitter when the trigger signal is a sine wave, again as in an oscilloscope circuit. Figure 9-14 shows a typical waveform generated by triggering.

Return the switch settings to normal retrace and self-triggering and remove the triggering input. Check the sine-wave output. With the values shown, it may show good resemblance to a sine wave, but probably will be distorted, most probably near the peak of the wave. This distortion can be partially compensated by adjustment of resistors R_A, R_B,

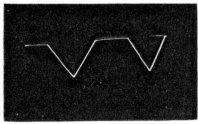

Fig. 9-14. Waveform available with trigger operation. Trigger period is approximately three times the calibrated period, giving 66% duty cycle of the triangular (and square) wave.

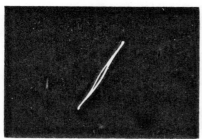

Fig. 9-15. Lissajous figure generated during adjustment of the sine-wave generator. Horizontal referene is 60 Hz power line, vertical is sine-wave terminal. Adjustment is complete when nearly perfect straight line is formed. Small hook at ends is typical of this class of sine-wave generators.

and R_C. To make this adjustment, feed the horizontal axis of your oscilloscope with a signal having good waveform, possibly from an audio oscillator or from the 60 Hz power line.

Feed the output signal of the function generator to the scope's vertical input and adjust the scope gain until an equal response is obtained on the two axes. Adjust the frequency of the function generator to the same frequency as the reference. What is wanted is a 1:1 Lissajous figure—a circle if the two signals are 90° out of phase, a straight line if they are in phase. Use the sync input of the function generator to obtain this, if necessary.

Inspection of the figure will show the area where the distortion is occuring; this is usually easiest to see when the two signals are in phase, giving nearly a straight line. Now adjust R_C for distortion near the peak output, R_B for distortion near the midvalue, and R_A for distortion near the axis, repeating until the closest approach to a straight line is obtained. It should be possible to get a very close approximation, probably with a small hook at the end. A typical approximation is shown in Fig. 9-15. If this wave is not obtained, or is not symmetrical, replace one or more of the switching diodes as needed. On a conventional sweep, the resulting wave should have a very close resemblance to a sine wave, the residual distortion showing as a small peak at the crest of the sine wave. Figures 9-16 and 9-17 show the sine generator output with normal and with fast retrace.

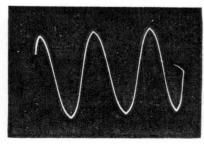

Fig. 9-16. Sine-wave output at low frequency. The small amount of residual distortion is at high frequency, and can be removed with a low-pass filter, if necessary.

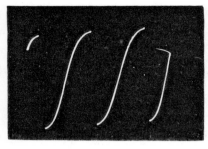

Fig. 9-17. Half-sine output available by switching to fast retrace. This is a fair approximation to a sine wave over the range from $-90°$ to $+90°$.

The final adjustment is needed only if the maximum possible upper limit of frequency is desired. This adjustment involves decreasing the stability margin of the op-amps, thereby increasing their upper frequency response and the switching speed. For this step, decrease the values of C4 and R10 until ringing is noticed on the highest speed sweep. Figure 9-18 shows the waveform obtained in the prototype by this step. With some op-amps it may be possible to also reduce the values of C1 and R4, or to eliminate compensation completely. With others some compensation will still be needed if oscillation is to be avoided.

CALIBRATION

For calibration, first place a dial with arbitrary markings on R7, as described in Chapter 3. Set the function generator for normal retrace and continous-wave operation, with no external trigger. Measure the frequency-versus-angle setting of the variable resistance with the selector switch set for a convenient frequency range. Record these in a form such as that in Table 9-2. Convert the readings of frequency to period, and then take half of this for the calibration value. Divide by the scale factor to get the dial markings.

With the values shown in the table, the dial markings run from 2 to 20. This range should be divided into nearly equal steps, approximately logarithmic ones if an audio-taper potentiometer is used, as in the prototype. Scale markings

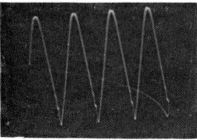

Fig. 9-18. Sawtooth wave output at minimum period. Op-amp compensation has been adjusted to allow noticeable ringing, the condition for maximum frequency response.

Table 9-2. Example of Dial Calibration.

	DIAL ANGLE	FREQ, Hz	PERIOD, μsec	HALF-PERIOD, μsec
MEASURED VALUES	0°	2487	402	201
		2137	479	249
	80°	1058	946	473
	120°	622	1610	805
	etc.	etc.	etc.	etc.

	HALF-PERIOD, μsec	DIAL	DIAL ANGLE	MULTIPLIER
VALVES FROM CURVE	200	2	35°	
	300	3	55°	
	400	4	72°	
	500	5		
			85°	$10^2 \mu sec$
	700	7	105°	
	10,000	10	137°	
	etc.	etc.	etc.	

selected for the prototype are evident in Fig. 9-1. There is room for closer calibration if desired—say, every half-unit from 2 to 10, and then each unit value from 10 to 20. Mark the dials as described in Chapter 2, then complete and install them.

USE OF THE INSTRUMENT

The most important use of the function generator is provision of very low-frequency signals, square waves, triangles, sawtooth and sine waves, and pulses.

In use, it is necessary to remember the output levels of the function generator, and also to remember the fact that the unit is intended for use only with high-impedance loads: the load resistance must not be less than about 5000Ω; otherwise, the waveform output of the function generator may be degraded.

The load and level limitation can often be avoided by using a simple L-pad attenuator at the output. This consists of two resistors—a high resistance in series with the lead from the function generator, and a low resistance in parallel with the input to the circuit receiving the function waveform. The shunt resistance can be chosen to match the required input of the instrument under test, and the series resistance chosen to give the desired voltage level. If variable levels are required, the series or parallel resistance may be made variable.

If this method of resistance padding is not satisfactory, the step-gain amplifier may be used to give a calibrated level of attenuation. However, even this amplifier is not intended to provide an appreciable amount of power. If a higher power level is needed—for example, to drive a loudspeaker—a power

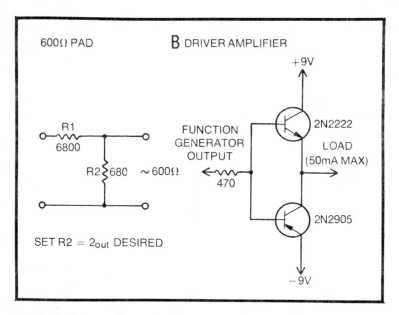

Fig. 9-19. Methods of providing output buffering: (A) resistance pad, giving 600Ω output with 20 dB attenuation; (B) amplifier capable of providing a few hundred milliamperes to low-impedance loads. (There is some crossover distortion with simple amplifier.)

amplifier should be provided. Figure 9-19 shows a simple push-pull amplifier capable of driving resistances of 5 - 10Ω. This amplifier does have a small amount of distortion near the zero-crossing point. If this is important, the output circuit of a good hi-fi amplifier can be used instead of this simplified circuit.

Sometimes special waveforms are needed—for example, a trapezoidal waveform to drive the deflection yoke of a videoscope or radar indicator. The step-gain amplifier can be used in the summing mode to add the outputs of the function generator. For very high flexibility of wave formation, two function generators are useful. By use of the synchronizing feature, these can be locked together and their respective outputs fed to the summing amplifier. Interposing the tone-burst generator in the sync circuit will allow very high flexibility of trigger delay, thereby extending the range of waveshapes which can be created. And, of course, simple use of the synchronizing and triggering feature will allow a number of special waveshapes to be created.

Another technique of wave generation is to use a combination of dividers and shaping elements based on diodes, just as is done in the sine-wave circuit of the function generator. Such a circuit may be built up separately, using variable resistors to give a variable-wave generator, with the desired wave being obtained by trial and error. Still another technique of generation of additonal waveforms is to use the step-gain amplifier as an integrator by connecting external capacitors from the output to the node terminal. This will give the integral of the input wave. For example, it will form a parabolic wave from a triangle, or a cosine wave from the sine wave. Differentiation can also be used by capacitive coupling from the function generator to the summing amplifier.

The function is a useful adjunct to an oscilloscope, and especially to the less expensive types which lack precision sweep circuits. The experimenter may want to construct a function generator in a case specifically designed for mounting on his oscilloscope, making permanent connections to the oscilloscope. Sweeps down to 1 sec/cm or even slower, plus a triggered sweep, can be provided by this method.

The function generator really comes into its own in analog computations. The experimenter may not recognize that he has built up, by now, the basics of an analog computation scheme. The combination of the function generator and the step-gain amplifier will permit solving the simpler problems—those involving only one integration or one summing. Additional step-gain amplifiers, possibly supplemented by the tone-burst generator, will provide an analog computation setup of great flexibility. The extensive literature on analog computation should be consulted for the type of problems for which this technique is suited, the equations and relations involved, and for the techniques of transferring from a problem to a specific connection of integrators, summers, etc.

In the general laboratory, the most likely use of a function generator is as an oscillator. It can also be used as a calibrated time-delay generator, by setting the controls for trigger operation, with pulse input to sync. Still another use is as a tracking filter that has a constant output. For this, set the selector to continuous-wave operation and feed the input to the sync terminal.

FUNCTION GENERATOR SPECIFICATIONS
- Square, triangular, sine output waveforms
- Pulse, sawtooth, half-sine outputs
- Frequency range from 0.02 Hz to 10 kHz.
- Synchronizable
- Trigger operation.

PARTS LIST FOR FUNCTION GENERATOR

C1 — 220 pF mica capacitor
C2 — 47 pF mica capacitor
C3 — 47 pF mica capacitor
C4 — 220 pF mica capacitor
C5 — 0.0002 mica or ceramic disc capacitor
C6 — 0.002 mica or ceramic disc capacitor
C7 — 0.02 mica or ceramic disc capacitor
C8 — 0.2 Mylar capacitor
C9 — 2.0 tantalum capacitor, 35V
C10— 20.0 tantalum capacitor, 35V
R1 — 100K All resistors ½W
R2 — 100K
R3 — 10K
R4 — 1500Ω
R5 — 100K
R6 — 91K
R7 — 1M potentiometer, audio or log taper (low backlash)
R8 — 1.0K
R9 — 10K
R10 — 1500Ω
RA — 100K
RB — 33K
RC — 15K (see text)
D1–D8 — Silicon 1N914, etc. (high front/back ratio)
IC1, 2 — μA704, TO-5

Chapter 10

The Q-Meter

The Q-meter is probably the most useful single instrument for RF work. As its name indicates, its basic purpose is to measure the figure of merit or Q-factor of inductors. It is also calibrated to read inductance directly and to measure capacitance by substitution, over the range of coil and capacitor sizes usually encountered in RF work. Because use is simple and fast, it is very valuable during design and construction, and quite useful in modification and repair. Typical uses are to determine the value of RF components, to make sure that a component to be installed is good, and to pretune coils. With a few auxiliaries, measurements of antennas and transmission lines can also be made. It is truly a multipurpose RF instrument. Figure 10-1 shows the miniature lab version of the Q-meter.

Fig. 10-1. The Q-meter, ready for disassembly. One of the early instruments designed in this series, the panel is finished with black lettering on satin-finish aluminum.

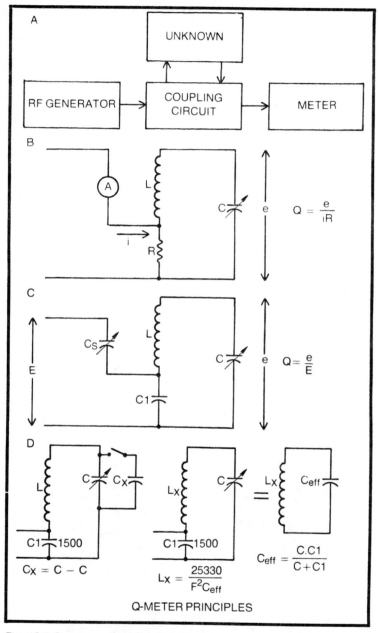

Fig. 10-2. Q-meter principles: (A) basic diagram of q-meter; (B) coupling circuit of commercial meters using current injection; (C) coupling circuit of commercial meters using current injection into resonant circuit; (D) measurement of unknown **C** and **L**.

BASIC THEORY

The principle of the Q-meter is shown in Fig. 10-2A. There is an RF generator, some circuitry to couple this RF energy to a resonant circuit (which includes the unknown component under test), and a metering circuit. Basically, the instrument measures the ratio of the energy input to the resonant circuit and the energy present in the circuit.

Laboratory-grade Q-meters usually measure the energy input by measuring the current passed through a known small RF resistance, and the energy present by measuring the voltage across the resonant circuit. The basic circuit and the Q relation of such a meter is shown in Fig. 10-2B.

This method is very satisfactory, but the cost of the RF current meter and the standard resistor is too high for use here. Therefore, an alternate method, shown in Fig. 10-2C is used. At the input, a capacitive divider (C_s and C1) is used to inject a fraction of a known voltage into the resonant circuit. The voltage across the circuit is also measured, the Q relation being as shown. (In practice, it is not necessary to determine the value of the factor k, since this is eliminated by the calibration process.) There is nothing new or novel about the circuit—for example, it was used at one time by Heathkit in a Q-meter which no longer appears in the company's catalog.

By calibrating capacitor C and knowing the frequency, the basis for measuring unknown inductances and capacitors is established. Figure 10-2D shows the technique. For inductance, several scale ranges can be provided, by operating at different frequencies. In this design the frequencies of 7900, 2500, 790 and 250 kHz are used, the last three to give scale multipliers of 10, 100, and 1000 times the basic calibration at 7900 kHz. The capacitance measurement can be made at any frequency; practically, the frequency is determined by the coil used to obtain resonance. Details of these measurements are described later.

DETAIL CIRCUIT

The complete circuit of the Q-meter is shown in Fig. 10-3. The RF generator is a transistor oscillator and emitter follower combination, with coil switching to cover the range of 200 kHz to 18 MHz. Except for the tuning range, this is basically the circuit of a transmitter variable-frequency oscillator (VFO). A jack is provided to allow the oscillator to be used for general lab purposes. Because it uses good high-Q toroids, shielding, and load isolation, its short-term stability is good, sufficient for use as a beat-frequency oscillator (BFO), for example.

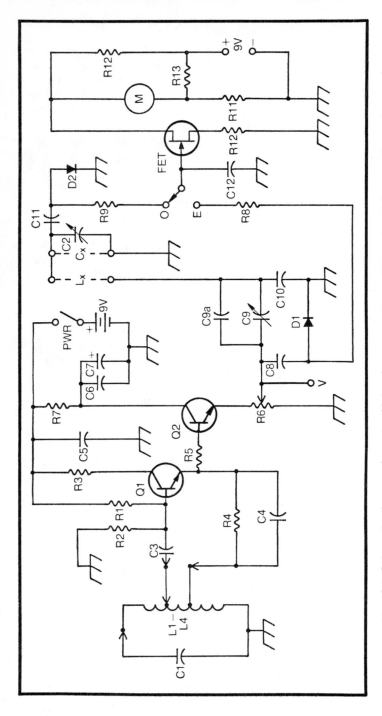

Fig. 10-3. Schematic diagram of Q-meter. Major elements are the oscillator circuit, the coupling and unknown section, and the meter circuit.

There are two voltmeters, which include the two shunt diode rectifiers, FET current amplifier, and necessary provisions for zero setting and switching. Again, there is nothing novel; the circuit has been used for such functions as S-meters, SWR meters, etc. The voltmeter is also available for general lab use, the terminals being the two terminals for unknown capacitors, with the meter switch in the Q position.

The pi-network which forms the heart of the measuring circuit includes one special component: a low-inductance capacitor used to inject the known voltage into the tuned circuit. It is made by removing the leads from a small ceramic disc capacitor. The larger the value of this unit, the more accurate the Q indication can be made. The value used here is a compromise, to favor simplicity in the current amplifier and to allow use of a low-cost meter. The other element of the capacitance divider is in two parts, an air trimmer (C9) and a fixed padder (C9a). Adjustment of these is part of the calibration procedure.

CONSTRUCTION

As shown in the photographs, the Q meter is built to the standard size of the miniature lab instruments. Most of the components are mounted on a PC board, which is laid out for low lead inductance in critical circuits. The board is supported by the terminals of the L_X and C_X binding posts, as a further way to keep stray inductance and capacitance low. Except for coils L1—L4, all components are panel mounted.

Board layout of the prototype is shown in Fig. 10-4. The areas marked "Mounting Pad" fasten to the panel binding posts. The special capacitor, C10, mounts in the small slot between two mounting pads.

Component placement and lead connections are shown in Fig. 10-5. The leads marked L_A, L_B go to the selector switch, which is a three-pole, four-position type. Coils connect directly to the switch, and are supported by a length of insulated bus bar, a length of spaghetti-insulated 14-gage solid wire fastened to capacitor C1 and to a large lug under the switch mounting. Figure 10-6 shows this mounting arrangement.

Panel layout of the prototype is shown in Fig. 10-7. There is sufficient space for several types of meters; the prototype uses a small surplus unit. More modern 1¼-inch square meters are available from several sources, and can be used, changing the size and location of the mounting hole to fit. If a larger meter is used, it may be necessary to move C2 and the

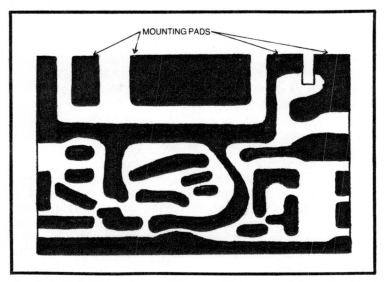

Fig. 10-4. Circuit board layout, 2:1 scale. Original board drawn freehand using an ink-resist pen.

terminals, which will require a change in board layout to make the mounting pads correspond.

In common with other units of the series, the Q-meter is designed for use with an external battery supply. Note that

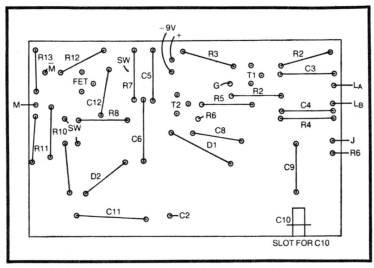

Fig. 10-5. Drilling template, parts location, and lead connection; scale=2:1. C10 is mounted at right angle to board, in a slot cut to just clear the unit.

Fig. 10-6. Internal view of Q-meter. Lead- and cord-supported toroidal coils at lower right. The shield attached to the left capacitor is to eliminate stray oscillator coupling into the measuring circuit.

there is no power switch—the battery must be disconnected to turn the unit off. (The schematic and parts list *do* show a power switch, however.)

Special Components

The only components requiring special work are capacitor C10 and the coils. The capacitor is made from a 0.0015 μF disc ceramic, an epoxy-covered unit approximately ⅜ inch in diameter. Select a unit within about 2% of specified capacitance, using the RC(L) bridge. The plastic of this may be loosened by soaking in fingernail-polish remover for twenty-four hours or so, and then lifting or scraping it free of the ceramic. The leads may be removed with a small soldering iron, leaving a disc, metalized on both sides (about ¼ inch diameter by 1/16 inch thick).

The coils (see Table 10-1) are wound on toroid cores. The low-frequency coil should be bank-wound, winding about five

Table 10-1. Coils For Q-Meter.

COIL	WINDING INSTRUCTIONS
L1	7 turns No.28E on CF-108 core, tapped 3.5, 7 turns. Range 6000 kHz – 17.1 MHz.
L2	24 turns No.28E on CF-108 core, tapped 6, 12 turns. Range 1950 – 7000 kHz.
L3	84 turns No.28E on CF-108 core, tapped 21, 42 turns. Range 540 – 2000 kHz.
L4	240 turns No.32E on CF-111 core, tapped 60, 120 turns. Range 160 – 620 kHz.
Indiana General Cores available from: Permage Corp. 88-06 Van Wyck Expressway Jamaica, N.Y. 11418	

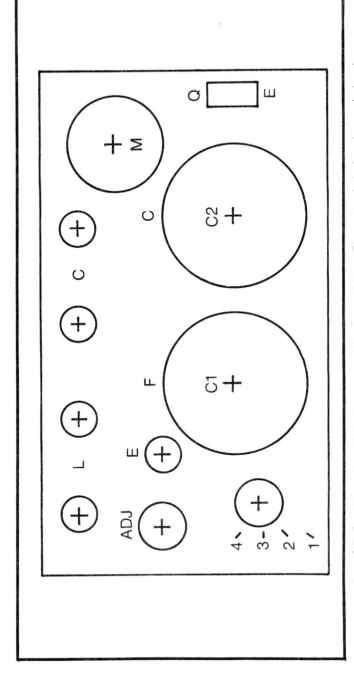

Fig. 10-7. Panel layout and lettering, as used in the prototype. The meter hole location and size, in particular, should be adjusted to the meter used. Improved appearance would be possible with finish as in Chapters 8 or 9.

turns then overlapping for about three layers. Other coils are layer-wound. Space the windings to fill the core on all coils.

Assembly Sequence

First, lay out and drill the panel, making sure that there is no interference with the capacitor rotors. Apply the panel finish desired and the lettering. Mount the panel-carried parts. Form the solder lugs on the binding posts into right-angle brackets, and mark their location on the foil side of the PC board. Lay out the board following Fig. 10-4, if necessary adjusting the mounting pad locations to fit. Mask and etch the board. Cut the slot for C10. Drill the solder lugs and the board for small screws. Make up a small aluminum shield to prevent inadvertent coupling from C1 to C2; the shape is evident when the capacitors are mounted. Make up the coil mounting bus and insulate it with spaghetti.

Mount the board components, temporarily omitting C9a and saving C10 for last. I prefer to use sockets for transistors and to mount rectifiers on the most accessible side of the board, usually the foil side. When mounting C10, first build up a deposit of solder on each side of the slot, insert the disc, and flow the solder to the disc.

After checking, mount the board to the binding post lugs with screws, then solder the lugs to the board. Mount the meter. Complete the wiring, and dress the RF leads for good clearance. Mount the coils to the switch, using nylon thread to fasten the coils to the bus wire.

This Q-meter was the first of the miniature lab units to be constructed. It was finished in bare aluminum with a minimum amount of lettering. It has been allowed to remain in the original condition, primarily to show the appearance improvements developed for later instruments, such as those of Chapters 7 and 8.

INITIAL CHECK AND ADJUSTMENT

Table 10-2 shows the values measured for the prototype by a resistance and voltage check *before the transistors were installed,* and a voltage check with transistors in place. Any marked departure from these values probably means a defective component or wiring error.

The first operating check is for parasitics. Use the widest-coverage receiver available to check this. A low-frequency parasitic will show up as a multiple signal— a *comb* spaced at the frequency of the parasitic. The most probable cure for this is a lower-gain transistor for Q1. A

Table 10-2. Check Readings.

| DEVICE | POINT | TRANSISTOR REMOVED | | TRANSISTOR IN PLACE |
		RESISTANCE	VOLTAGE	VOLTAGE
Q1	Base	15K	1.8	2.0
	Emitter	240	0	2.0
	Collector	5K	8.7	8.0
Q2	Base	1800	0	2.0
	Emitter	1200	0	3.4
	Collector	5K	8.7	8.1
FET	Gate	3M	0	0.1
	Source	240	0	0.8
	Drain	5K	8.5	7.8
	E_{bb}	—	8.7	8.7
Voltage measurements made at 30.000 Ω/V, 10V scale.				

partially open capacitor at C5, C6, or C7 may cause this. A high-frequency parasitic may show up in the VHF range; it might appear as "roughness" on the fundamental. Corrections include a transistor of lower cutoff frequency, adjustment of lead position, and possibly installation of ferrite beads on the leads to the coil switch and to C1. In the prototype there was no sign of VHF parasitics with any transistor tried; a low-frequency parasitic did originally appear with some transistors and was corrected by reducing C3 and C4 to their values.

The next step is to set the zero on the meter. Set R6 to the ground end, and the meter switch to Q. Shunt R10 with an external variable resistor, and adjust it until the meter reads zero. Measure the value of the two resistors in parallel (with the FET out of its socket), and replace R10 with this value (a variable resistor could be used for R10). Check that the zero reading does not change appreciably with the meter switch on E: a small change will be calibrated out. If the change is large, make sure that the arm of R6 actually reaches ground, and double-check the diodes and coupling capacitors. Also, try moving the RF leads to see if there is unwanted coupling.

Temporarily place a broadcast-band coil of about 100 μH across the L_x terminals, set C1 to midscale, and set C9 to maximum capacitance. Set the *frequency* switch to *range 3* (the broadcast band) and the *meter* switch to E. Adjust R6 for full-scale meter reading. Throw the meter switch to E and vary C2. At some point the meter reading should reach a maximum, indicating resonance. The deflection will depend on the coil quality, and is probably less than one-quarter scale.

It is a good idea to check the waveform of the oscillator signal, especially on the lower bands. Distortion, as in Fig. 10-8, can cause false resonances at submultiples of the true

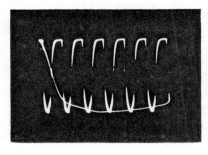

Fig. 10-8. Low frequency of oscillator. The harmonic distortion may cause a false reading when high-Q, high-inductance coils are being measured.

resonant frequency. The easiest correction is to change the oscillator transistor to one of lower beta.

With these checks complete, the Q-meter should be functional and ready for calibration.

CALIBRATION

If an electronics laboratory is available, the best method of calibration is to use the standards available—it's easier and more accurate. The following method assumes that the only auxiliary equipment available is a radio receiver covering at least the range 550 kHz to 18 MHz. The accuracy of the calibration will not be as good as if a full set of standards is available, but it is adequate for most practical purposes.

The first calibration is for oscillator frequency. Use the most accurate method you can—the dial calibration of the radio receiver if necessary; beating against a crystal standard is better, and measurement with the digital frequency meter best of all.

For the rest of the calibration steps it is necessary to construct a working standard for L, which also serves as the standard for Q. A second standard, for C, is not absolutely necessary, but makes a good method of checking accuracy. The construction of these two standards is shown in Fig. 10-9, which also shows the method of checking coil accuracy.

To prepare the capacitance and inductance calibration of C2, plug in the standard inductor, set the frequency successively to the values tabulated in Table 10-3, and record the dial readings of C2 which give resonance. You'll be able to see the basis of this calibration by noting (in Fig. 10-2B) that resonance occurs as a result of L_X, C, and C1 in series. As a result, there are two capacitances of interest, the actual value of C, and the effective value of C and C1 in series (C_{EFF}). Capacitors C and C1 of Fig. 10-2E are the same as C2 and C10 of Fig. 10-3, the schematic. For a constant frequency, there is a particular value of inductance which will resonate at a given

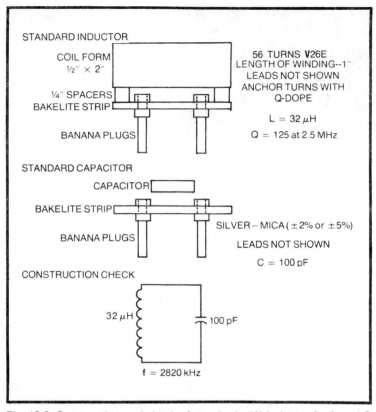

Fig. 10-9. Construction and check of standards. (A) Inductor for **L** and **Q** calibration. Follow dimensions closely. (B) Typical capacitor for calibration (see Fig. 3-3 for others which may be used). (C) Check of coil accuracy, using standard capacitor.

value of C2, which is the basis for the inductance calibration. Table 10-3 gives a selection of round-number values of L, C_{EFF}, and C2, the corresponding resonant frequencies of C2 and the 32 μH standard, and has blanks for noting the dial readings of C2. This table is calculated for C10 equal to 1500 pF, and should be recalculated for the *measured* value if appreciably different.

The Q calibration is made last and is based on the fact that the standard of Fig. 10-9 has a Q of 125 at 2.5 MHz. Set the frequency at 2.5 MHz, R6 to give 0.8 mA deflection with the meter switch on E, and plug in the standard inductor. Switch to Q and tune for maximum deflection. Adjust C9 for a meter reading of 0.63 mA, adding capacitance with C9a as needed—probably 15—20 pF. This gives a Q reading of 200 full

Table 10-3. Frequencies for Calibration, $L = 32\ \mu H$.

CALIBRATION	SCALE VALUE	FREQUENCY, kHz	DIALING READING
C2	25	5680	
	50	4050	
	100	2920	
	150	2390	
	200	2120	
	250	1880	
	300	1780	
	350	1670	
C_{EFF}	25	5650	
	50	3980	
	100	2820	
	150	2300	
	200	1980	
	250	1780	
L	1	1400	
	1.5	1710	
	2	1970	
	3	2420	
	5	3120	
	8	3950	
	10	4420	
	15	5410	
	20	6250	

scale. The scale will be reasonably linear down to a Q of about 20.

Calibration of the Q scale can be accomplished by measuring the voltage across the L_X terminals with an RF probe of an electronic voltmeter. Adjust C2 as needed to maintain resonance, and vary R6 to obtain various values of voltage. Record the Q-meter scale reading versus VTVM voltage reading. The ratio of measured voltage at any meter reading to the voltage at a meter reading of 0.63 mA, times 125, is the Q calibration.

In the prototype, calibration scales are fastened to the side of the meter as shown in Fig. 10-10, using graph paper protected by plastic film. A better appearance could be obtained by painting the case sides white, and marking the scale and calibration on this. Alternatively, the calibration can be kept in the lab notebook.

There is sufficient space on the bottom half of the miniature dials to allow direct calibration of one scale for L and another for C2. A separate graph of C_{EFF} versus dial reading is adequate, since this scale is not used much. For the

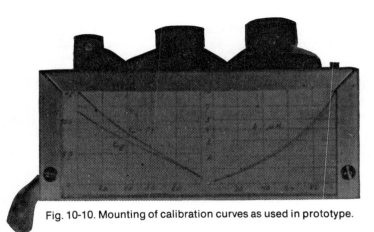

Fig. 10-10. Mounting of calibration curves as used in prototype.

frequency calibration, it is suggested that the points for 250, 790, 2500, and 7900 kHz be marked on the dial, and the full calibration be a graph mounted on the side of the instrument. As an alternative, a special dial scale for C1 could be made up.

USE OF THE Q-METER

The following is a reasonably complete list of the measurements which can be made with the Q-meter.

Unknown L

Set F to one of the frequencies, 250, 790, 2500, 7900 kHz, and connect L to the L_X terminals. Vary C2 for resonance, changing frequency if necessary. Adjust R6 to give the correct E indication of 0.8 mA, read L on C2 dial, with the correct multiplier, and read Q on the meter. If the meter reading is off scale, reduce input voltage E by factor of 2 or 4 (*i.e.*, E meter reading equals 0.4 or 0.2 mA), and multiply Q by this factor. The inductance scale can be multiplied by ½ or 2 by setting the frequency to 1.4 or to 0.7 the normal frequency.

Unknown C

Plug the standard or any convenient coil into the L_X terminals and set C2 to a convenient value—say, 300 pF. Adjust C1 for resonance. Connect the unknown capacitor across the C_X terminals and retune C2 for resonance. The difference between C2 readings is the unknown capacitance. If there is a measurable difference in the Q reading for the two conditions, the unknown capacitor is defective.

Unknown f

Use the instrument as a wavemeter by plugging a suitable coil into the L_X terminals. Set R6 to zero E output and switch to

154

the Q position: the FET meter becomes the indicator. Vary C2 for maximum reading. Subsequently, the internal oscillator can be used to determine the resonant frequency of L and C2, which is the approximate frequency.

Field-Strength Indication

As for a wavemeter, connect a length of wire to the terminal which is common to both L_X and C_X, to serve as an antenna. Read relative field strength on meter.

Distributed C and True L (of Coil)

Measure C2 at a frequency, f_1, and C2′ at twice this frequency, f_2. Then

$$C_{dist} = \frac{C2 - 4C2'}{3} \qquad L_{true} = \frac{19000}{f_1^2(C2 - C2')}$$

Extended Range of L

For small coils, add known capacitors across C2 in steps of about 300 pF. The Q measurement becomes progressively less correct as C is added. For high inductance coils, use the *general unknown* method below.

Extend Range of C

Note, from Fig. 10-2E, that L_X C and C1 form a series-resonant circuit. The resonant frequency changes with change in C (C2 in schematic). To use the method, place the standard coil at the inductance terminals, set C2 to a convenient value—say, 100 pF—and connect the unknown to C_X. Determine the frequency of resonance. For this frequency, calculate first the *effective* capacitance, then the value of the unknown. It is convenient to precalculate these values, making a plot of C versus f. The method is useful for values of C from 0 to about 1500 pF.

General Unknowns

For antennas, choke coils, etc., measure C2 and Q of a good-quality coil, and C2′, Q′ with the unknown in parallel with the coil. Then

$$C = C2 - C2'$$

$$B = 2\ fC$$

$$\frac{1}{Q} = \frac{B}{G} = \frac{C2}{C} \quad \frac{1000}{\dfrac{1000}{Q} - \dfrac{1000'}{Q}}$$

$$\frac{1}{Z} = Y = G + jB$$

If C2′ is greater than C2, the unknown is inductive. For low impedance or highly reactive unknowns, connect 100 pF in series with the unknown, and calculate first the total series impedance, then the unknown, assuming that the added capacitor is loss-free.

AUXILIARIES

The usefulness and speed of operation of the Q-meter can be increased if some auxiliary components are made up. Suggested ones are:

- Alligator clips with attached banana plugs for fast connection of unknowns.
- Extra standard coils and capacitors: 2μH, 200 μH, 2 mH, plus 100 pF and 300 pF.
- Standard coils for Z measurement, with built-in 100 pF series capacitor.
- Transistor—diode capacitance tester, with built-in isolating chokes and bias circuitry.
- Telescoping whip antenna for field-strength measurements.

THE Q-METER SPECIFICATIONS

- Measures L in 4 ranges, 1 μH to 10 mH
- Measures C in 2 ranges, 0 to 1500 pF
- Measures Q in 3 ranges, 0 to 200, with $\times 1$, $\times 2$ or $\times 4$ multiplier.
- Also serves as:

  —General laboratory oscillator, 200 kHz — 18 MHz
  —Sensitive RF voltmeter
  —Wavemeter
  —Field-strength meter
  —Usable for measuring any RF impedance

PARTS LIST FOR Q METER

C1, C2	—	350 pF miniature air variable
C3, C4	—	0.0015 ceramic
C5, C6	—	0.1 uF Mylar
C7	—	10 uF electrolytic
C8, C11	—	0.001 ceramic
C9	—	3-15 pF air trimmer
C9a	—	15 pF ceramic
C10	—	0.0015 ceramic

C12	— 0.01 ceramic
R1	— 56K
R2, R4	— 15K
R3, R7	— 100Ω
R5	— 1K
R6	— 500 carbon variable (AB or equal)
R8, R9	— 3.3M
R10	— 390Ω
R11	— 4700Ω
R12, R13	— 220Ω
PWR	— SPST Switch
Q1, 2	— 2N706, etc. ($h_{FE} = 30$, $f_T = 200$ MHz)
FET	— Motorola MPF 103, etc. (general-purpose audio)
D1, 2	— 1N64A, etc. (high back resistance)
M	— 0–1 mA, 1″ diameter, International
J	— Phono Jack
Terminals	Binding posts, 5-way

Chapter 11
The Digital/Frequency Period Meter

Although counting techniques have been used in some of the instruments already described, the digital frequency meter is the only digital instrument in this miniature lab series. A frequency meter of the counting type is a good introduction to digital techniques. It is quite easy to work with, since a large number of RF cycles are involved in frequency measurements: counting or digital design technique lends itself especially well to this.

Actually the term *frequency meter* doesn't fully describe the wide range of usefulness of instruments of this type. The earlier name, EPUT meter (for *events per unit time*) was a much better choice. This class of instrument can measure any repetitive countable signal, or any occurrence which can be converted to a countable signal. Examples are the number of automobiles passing in unit time, or the number of disintegrations of a radioactive source, or the speed of an electric motor, etc. It is not necessary that the events be uniformly spaced; the only requirement is that the spacing between events exceed some minimum value if error is to be avoided.

Additionally, if a little care is taken, the meter can be inverted in action—it can become a *time per unit event* meter. This allows it to measure such quantities as the period of a wave, or the time intervals between cars passing, or the length of time to fill a tank with fluid, etc.

The digital frequency meter is a popular instrument, as it deserves to be.

BASIC PURPOSE OF INSTRUMENT

In design of this instrument two general goals were established. One was to fit a frequency meter into the standard

$2 \times 3 \times 5$ inch case size of miniature laboratory. The second goal was that the cost should be low—a maximum of $50 was set, and a goal of $20 per unit was kept in mind.

Neither of these goals is particularly easy. The size goal amounted to about 10% of the volume of commercial equipment available at the time the design was undertaken. The cost goal was less than one-third the price of kit equipment and about 10% the price of the commercial equipment available.

Obviously, to meet such cost and value goals, some features had to be sacrificed. On the other hand, some features were considered to be basic and to be required in the design. The most important of these were;

- Simple reading
- Usable to at least 10 MHz
- Resolution to 1 Hz
- Automatic recycling
- Manual override
- Readings of events *or* time.

It proved possible to obtain these features, plus a few additional ones, within the limits of volume and costs. Figure 11-1 is a photograph of the instrument.

CIRCUIT AND THEORY OF OPERATION

The minimum elements required in a frequency meter are shown in block form in Fig. 11-2. Countable signals must be in a

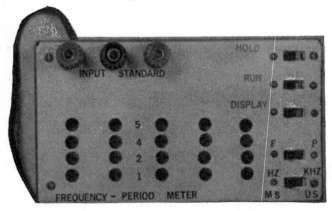

Fig. 11-1. The digital frequency/period meter. Reads frequency from 1 Hz to approximately 15 MHz, periods down to 1 μsec, by illuminated LEDs behind panel holes. Panel finished with black lettering on white background. As a result of frequent use, paint chipping is noticeable at switch opening corners. It can be prevented by using a good primer as the first paint coat.

160

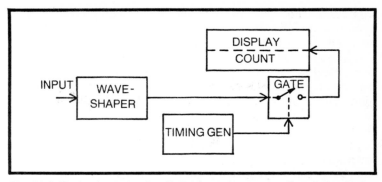

Fig. 11-2. Basic elements of a digital frequency meter. Resolution is determined by number of digits in display. Accuracy and range are established by timing generator.

form suitable for digital operation; basically, they must have fast rise time and be free of noise. These functions are taken care of in the input waveshaper. From the resulting wave train, the number of cycles within a specific time period must be selected and passed on for counting. This is done by the gate, which is controlled by the timing generator, or clock. Finally, the number of events within the gate is counted, and the count displayed in a form compatible with the timebase.

Commercial frequency meters use a display based on the number 10—in other words a display of decimal digits. The display may be by a special neon tube, incandescent lamps, or light-emitting diodes, but the form of the display is designed to resemble the standard number series, 0 through 9. In modern displays, t9 eliminate flicker and constantly changing display value, there is a sample-and-hold circuit, with means to update the display periodically.

Even in a simplified design, it seems necessary to retain the digital feature and the sample-and-hold function. However, in the usual design, each digit requires not only a counter, but the sample-and-hold circuit, conversion from the number system of the counter to the digital number of the display, and then the display itself. Each digit may require as many as four IC packages, although a single package combining these four functions is available. Unfortunately, the single-package unit is expensive; even the four-package system, for four digits, bought as surplus, would exceed the cost goal of the entire instrument. The conclusion is that a small inexpensive frequency meter must use a simplified display system.

One way of accomplishing this is to base the display on the number system used in the counter. Using counters readily available in surplus, two options are available. As usually connected, the counters count in binary, with internal reset, commonly called BCD, for *binary-coded decimal*. Standard counters can also be rearranged to count in a number system based on five and two, or *biquinary*. Displays based on these two counting possibilities are illustrated in Fig. 11-3. The first part shows the value or *weight* of a position, and the second part shows an example display. Neither of these displays resemble decimal digits. However, it takes about 30 seconds to learn the code, as shown by the examples on the right. Initially, in use, it takes perhaps one second per digit to mentally translate the code into a number. With a few hours of use, the pattern is read at a glance.

Considerable simplification is possible if the display is driven directly rather than through auxiliary circuits. This is now easy, by use of light-emitting diodes. Using one LED for each binary digit, or four per decimal digit, the patterns of Fig. 11-3 are readily developed. The small LEDs available require about 0.5 to 1V at 10–20 mA to provide good illumination. Since counters can supply as much as 50 mA of current, and have a *high* logic level of 2.5V minimum, they can drive the LEDs directly.

It must be remembered that each counter element must drive a successive stage: the counter cannot be loaded fully, or erratic operation will result. The loading can be avoided by turning off the display during the actual count period. This can also serve the function of the sample-and-hold feature, by

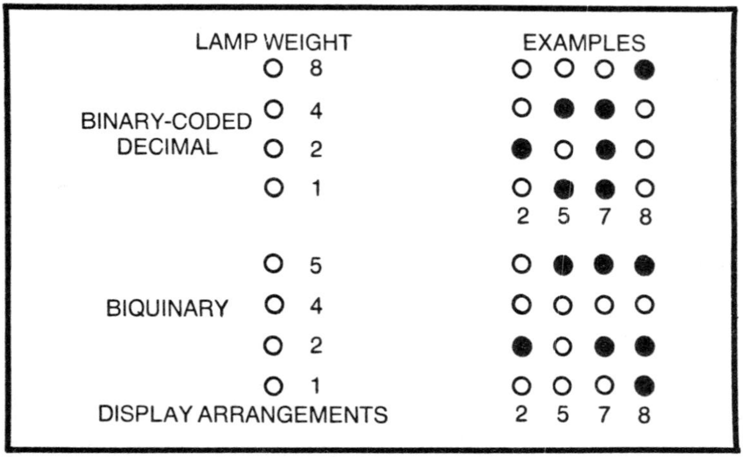

Fig. 11-3. Use of lamp positions to show measured values:
Relative weight of vertical position (left) and example (right) of 4-digit number. For BCD, count by twos, and reset on ten. For biquinary, count by fives, twice.

feeding the input pulses into the counter for the gate period, with the display off, then activating the display for a period long enough to allow the display to be read.

Of the many ways to accomplish these functions, Fig. 11-4 shows the principles of the method adopted. This was used because it is both simple and flexible. To start, assume that a chain of one-per-second pulses from the clock is numbered 0 through 9; the end of one chain and the start of another are shown. The first stage of a BCD counter divides by two, going *high* on odd pulses and *low* on even. The last stage of this counter goes high on the eighth pulse and is internally reset to low at the end of the tenth pulse, to start the new chain. By ANDing the 2 and 8 counts, a pulse length of one clock period is obtained, repeated each tenth period. This one-period pulse is the gate control signal. Before the gate turns on, the counter chain must be reset to zero. Also, the display should be turned off to avoid loading the counter during the count period. An easy way to accomplish this is to take the negation (*not* function of the eight level pulses, which gives a signal that is *high* for eight counts, then *low* for two counts. The *high* can be

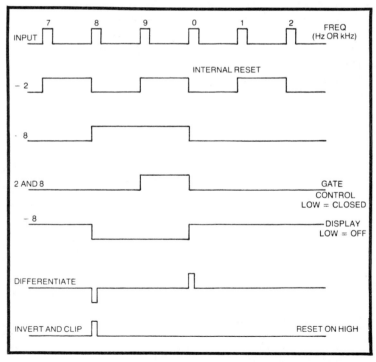

Fig. 11-4. Control pulse generation. (See text for formation and use.)

used to turn on the display. Also, this signal can be differentiated then clipped and inverted to form a pulse at its leading edge—this gives the *reset* signal.

The sequence gives one clock period for counting and eight clock periods for display, followed by one clock period for reset.

The circuitry to accomplish these functions is shown in Fig. 11-5. The control pulses, which may be for frequency or period, as described later, are fed to a standard 7490 digital counter. The 2 and 8 counts from this are fed to half of a 7413 Schmitt-trigger AND gate. The unknown is also fed to this gate. One input to the Schmitt circuit is connected to +5V so that it is always *high*. The 7413 performs the dual function of combining the 2 and 8 counts to give the gate, and of gating the unknown pulses into a stream of one count period *on*, eight count periods *off*.

Separately, the 8 output of the 7490 is inverted by one-fourth of a 7400 used as an inverter, to form the display gate. This is fed to another section of the same 7400 through a small coupling capacitor, which gives the differentiated pulses. The second half of the 7400 performs the functions of inverting and clipping, to provide the reset signal.

The 7490 counter has internal gates to allow it to be reset to zero, or to be preset to the 9 count. Referring back to Fig. 11-4, the 9 count is equivalent to allowing the gate to run continuously. Simply opening the number 6 or 7 lead of the 7490 (or both) allows the unknown to be fed to the count display

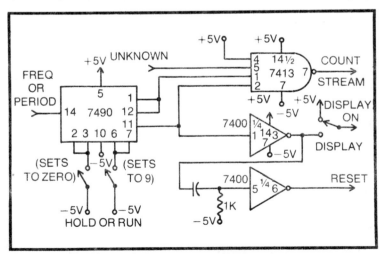

Fig. 11-5. Schematic of generator for gate and control pulses. This provides the pulses shown in Fig. 11-4. The 7413 is the gate.

circuit continuously. The switch to accomplish this is marked RUN in the open position. In a similar way, from Fig. 11-4, the result of setting the 7490 counter to zero at any time during the interval corresponding to pulses 0 to 8 is to block the counter reset. This causes the display to continue indefinitely, showing the last count accumulated. This action is obtained by opening the number 2 or 3 lead of the 7490 (or both). The switch to accomplish this is marked HOLD in the open position. Note that there is a chance of error if this switch is opened during the eighth interval, but this is evident by the fact that the display becomes zero. There is also a chance of error if the switch is opened during the ninth interval, which would cause a partial count to be accumulated. However, the risk of such an error is only 1 in 10, and the error is easily detected by the sudden change in reading from the preceding value. It did not seem worthwhile to add circuitry to avoid this error.

One additional function is easily provided. Sometimes it is desirable to have the display on continuously, even during the count period. This is obtained by switching the display control from its 7400 output to +5V. The switch to accomplish this is marked DISPLAY on the 5V position.

To meet size and cost goals, only five digits are used in the counter display chain, which allows a maximum count of 99,999. If the gate pulse is one second long, the maximum frequency which can be read without ambiguity is 99.999 kHz, readable to 1 Hz. To read higher frequencies without ambiguity, the gate can be shortened. The most useful is a 1 msec gate, which gives, in principle, a maximum capability of 99.999 MHz, readable to 1 kHz. However, the maximum capability of the 7490 counters is in the range of 15−25 MHz, which becomes the upper frequency limit of the counter. A prescaler must be used for higher frequencies. Actually, the number of digits could be decreased to four, while still obtaining full count capability; however, the overlap provided by the extra digit simplifies reading.

The circuit used for the count display system is shown in Fig. 11-6. Five cascaded 7490s form the counter. For each of these counter stages, the 1, 2, 4, and 8 level outputs are fed to LEDs to form the display, as in Fig. 11-3. Either of the display codes of Fig. 11-3 can be used, by arranging the connections of the LEDs and connecting the preceding stage output to input A of the 7490 for BCD (or to input BD for biquinary). Biquinary is used in the prototype, since the value seems to be slightly easier to read rapidly.

To prevent loading of the counter during the count accumulation time, the LEDs are inactivated by using a series

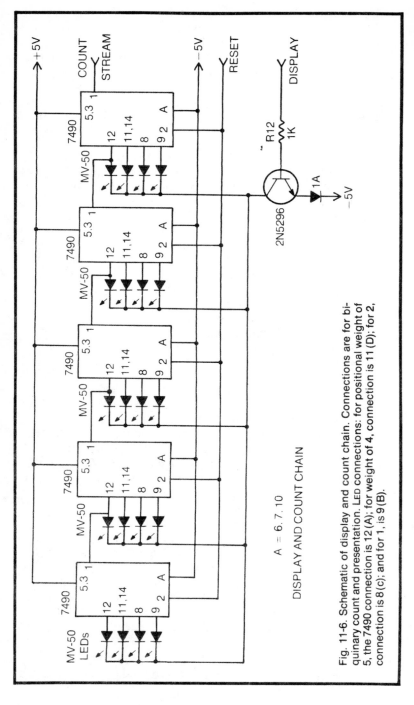

Fig. 11-6. Schematic of display and count chain. Connections are for bi-quinary count and presentation. LED connections: for positional weight of 5, the 7490 connection is 12 (A); for weight of 4, connection is 11 (D); for 2, connection is 8 (c); and for 1, is 9 (B).

A = 6.7.10

DISPLAY AND COUNT CHAIN

transistor as a switch. The control for this is the negated sign of the 8 count.

High levels of LED current give better visibility but greater heating in the 7490s, and more loading when the display is òn. The current is set by the characteristics of the 7490, the drop across the series gating transistor, and by the LED characteristics. Current control can be provided by connecting one or more series diodes in the emitter lead of the gate transistor. For the MV-50 LEDs used in the prototype, two series diodes gave weak illumination, while omitting the diodes gave excessive loading of the counters; one additional emitter diode gave a good compromise between high output and loading, and was used.

It is necessary that the clock have good frequency stability. This is obtained in this design by the use of an oscillator using an AT-cut crystal, specified for zero temperature coefficient. Minimum frequency for these crystals is about 1 MHz, which requires six stages of countdown to reach the 1 pps repetition rate needed. While the number of stages could be reduced by using low-frequency crystals, the AT cut gives better temperature stability and allows use of the simple multivibrator oscillator of Fig. 11-7. The multivibrator output feeds a Schmitt trigger, which is provided with input and output buffers to prevent loading changes. The counter chain, also shown by Fig. 11-7, is straightforward. In the prototype only three points of the chain are brought out: 1 MHz, the 1 kHz, and 1 Hz. These three are required for the counting arrangement used, as described below. Additionally, the MHz output is fed to a panel terminal, for incidental laboratory use.

The input circuit of the counter is shown in Fig. 11-8. This provides a high-impedance input, overvoltage protection, gain, and waveshaping. Direct coupling is used, which allows the counter to be used as an interval meter. The amplifier chain is relatively low gain—about 1V is required for reliable operation. The FET gate is protected by two transistors connected as zeners; these may be omitted if an internally protected FET is used. The output stage is the second half of the 7413 Schmitt AND gate already used. It is used as a simple Schmitt trigger, three inputs being connected to +5V.

The interconnections between these counter elements are shown in Fig. 11-9. Switching is arranged so that the gate can be fed by the input unknown for *frequency* measurement, or it can be fed by the clock for *period* measurement. The same switch also selects the gate control input. This is the clock when frequency is being measured, and the unknown when

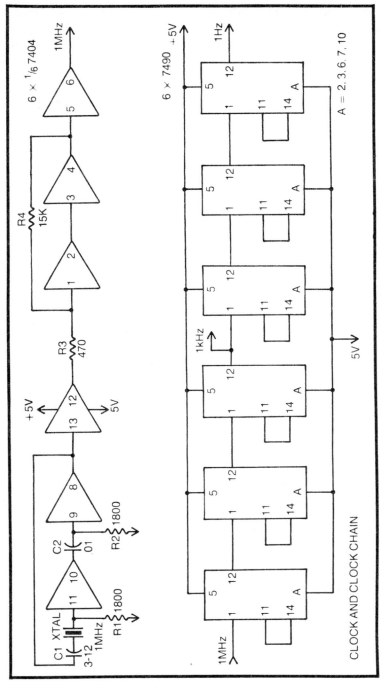

Fig. 11-7. Schematics of timing elements: (A) crystal-controlled clock; (B) clock chain (to divide 1 MHz clock pulses to 1 kHz and 1 Hz).

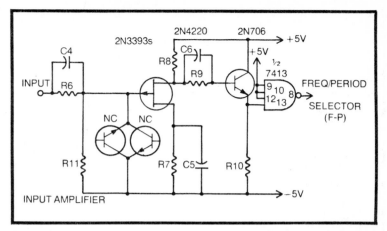

Fig. 11-8. Input amplifier and protective circuit. The 2N3393 transistors are used as zener protective diodes and can be omitted if a gate-protected FET is used.

period is being measured. A single clock switch is provided, to give the scales of hertz and kilohertz for frequency measurement, and the scale of microseconds or milliseconds for period measurement. If desired, the period scale can be changed to milliseconds and seconds by connection to the 1 kHz and the 1 Hz points.

The counter requires +5V for operation. The current drain is quite large, and varies by a considerable amount as the count cycle progresses. If AC line operation is to be used, *a regulated power supply is a necessity*. Battery operation is feasible: the minimum recommended source is a 6V lantern battery, with a series 1A silicon diode to give 5.4V to the counter. The silicon diode also provides polarity protection, a desirable feature. Because of the size of the battery, and the fact that the counter is quite crowded, a separate power pack is needed. The common LM309K regulator is satisfactory if an AC supply is used, or if +12V operation is needed. Don't forget to shunt the input and output of this by 0.01 μF disc ceramic capacitors with short leads, to prevent parasitic oscillations.

COMPONENTS

The integrated circuits used in this counter are available from a number of surplus houses at low cost—typically less than a dollar each. Most houses also carry the LEDs at low cost. The type of transistors and FETs is not critical—random samples from several surplus kit units were tried in the prototype and without exception gave satisfactory operation.

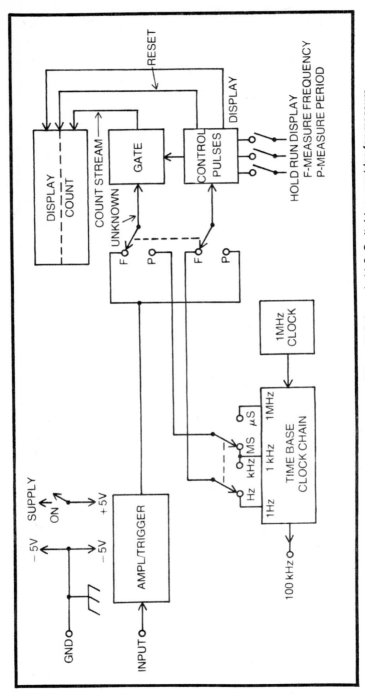

Fig. 11-9. Interconnections for the elements of Figs. 11-5 through 11-8. Switching provides for measurement of frequency and period, with two ranges for each. The three additional switches are detailed in Fig. 11-5.

The transistors used as protective zeners should be checked for porper operation before installation, especially if other than the specified type is used. The final chapter describes this check.

If the specified FET is used, the customary precautions against static electricity are needed. The LEDs are easily damaged by excessive heat; use heatsinks when soldering these in place.

Other than these points, there are no critical elements in the construction of the frequency meter.

GENERAL CONSTRUCTION

The general construction of the digital frequency meter is evident from Fig. 11-10. All components except terminals and switches are mounted on two PC boards. The board closest to the front panel carries the count accumulator and the display circuitry of Fig. 11-6. It is spaced one-half inch from the panel and mounted so the digit diodes are visible through holes in the panel. The separation between the PC board and the panel does result in some restriction in viewing angle, but the shadowing makes it easier to read the digit count under bright ambient lighting conditions.

The second board carries the remainder of the circuitry, that of Figs. 11-5, 11-7, and 11-8. In the prototype, the second

Fig. 11-10. Internal view of prototype. Mounting the two boards to the front panel section by standoffs is probably superior to the mounting shown. Because of crowding, an external power supply is necessary.

171

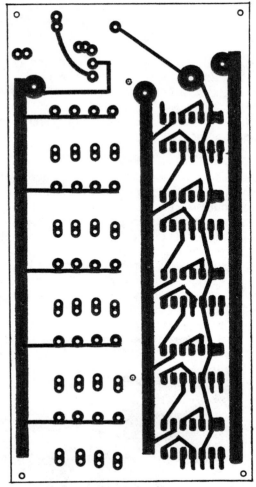

Fig. 11-11. Circuit board layout for the display/count section. Art shown was prepared with Bishop Graphics and "Easy-Etch" transfers, but original board was drawn freehand with ink-resist pen.

board is mounted on the back half of the minibox used for the cabinet; it might be better to mount both boards on the front panel, thereby avoiding lead flexing when the case is opened.

The interconnections and switching follow from Fig. 11-9. The layout of the two boards is shown in Figs. 11-11 and 11-12. The prototype boards were hand drawn, using an ink-resist pen. This is very tedious, in view of the number of integrated circuits involved, but is quite feasible. If facilities are available, it is recommended that the board be fabricated using photosensitive resist, using negatives made from these

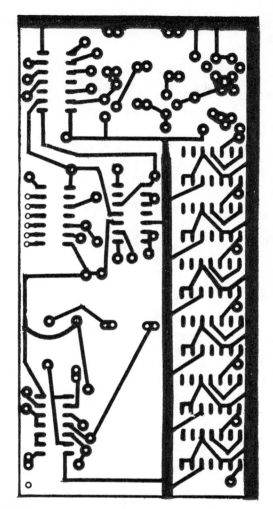

Fig. 11-12. Circuit board layout for other sections. While freehand masking is feasible, photocopy of "Easy-Etch" is suggested.

two figures. As with all other PC masks in the series, the art is shown here full-size, and may be reproduced photographically without corrections.

Board drilling and component placement are shown in Figs. 11-13 and 11-14. Four methods of mounting the IC devices are available. They can be soldered in directly, which is the least expensive and the least satisfactory method. Circuit-Stik is the second alternative. With this method, you apply peel-off adhesive "socket" forms directly to the board, then solder the

173

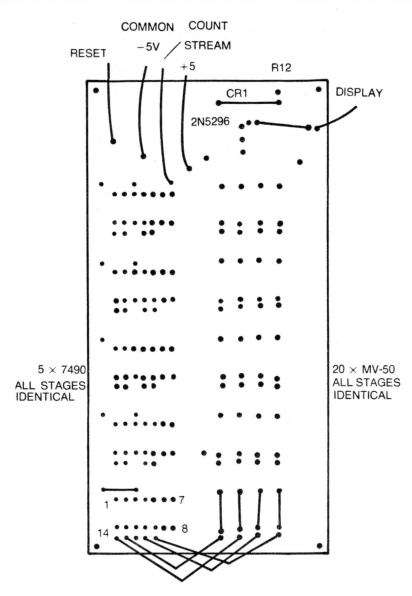

Fig. 11-13. Drilling template, parts location, and leads and jumpers for the display section. Jumpers and LEDs shown only for one decade.

ICs in place. For plug-in capability, use Molex terminals or commercial DIP sockets, the last being the most expensive course to follow. Molex terminals were used in the prototype as a cost-saving factor. Sockets are probably better in the long

174

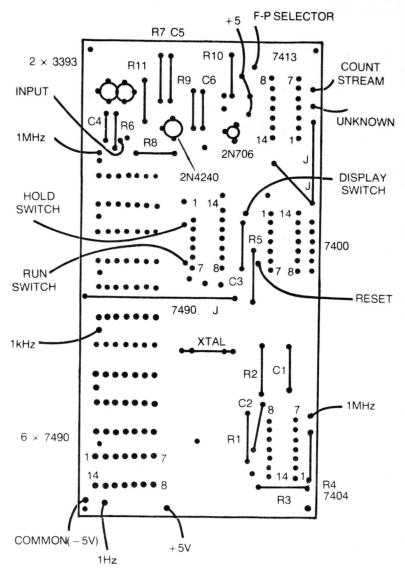

Fig. 11-14. Drilling template, parts location, and leads and jumpers for the second board.

run, especially when surplus integrated circuits are used, which may have to be changed.

If the Molex terminals are to be used, mount all other components first. When ready for the terminals, break off a set

175

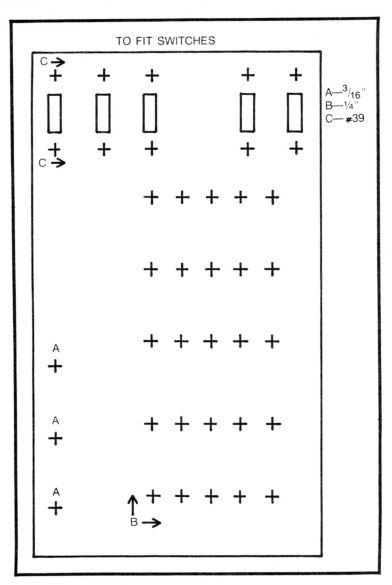

Fig. 11-15. Panel layout for the digital frequency meter. The wider spacing between the second and third set of LED viewing holes is intentional, to mark the location of the 1000s.

of seven from the roll in which they are supplied. Straighten the strip carefully, then insert the pins in the socket holes and solder it in place, making sure that the terminals are perpendicular to the board.

Inspect your work carefully. Leave the strip which holds the individual terminals together in place until all board work is done, then break off the strip by flexing with pliers. Practice on a piece of scrap board before starting the final unit.

The prototype originally used the same type of Molex terminals for transistor mounting. This was not very satisfactory; use of transistor sockets is recommended. The crystal socket can be made from the pins from an old tube socket—Loktal for miniature crystals, octal for FT-243 surplus units. Bend the socket to mount the crystal case parallel to the circuit board. The crystal case should be tied to the board with nylon cord.

The panel layout is shown in Fig. 11-15, and is also evident from Fig. 11-1. Two of the switch mounting screws also serve for board mounting. Alternate mounting points at the center of the boards are shown in Figs. 11-12 and 11-13, and can be used instead of the four-point mounting of the prototype.

The prototype used single-strand flexible wire for interconnection between boards and for connection to switches and terminals on the panel. It is suggested that this be used for initial testing, with the boards outside of the case. After everything is working properly, the final assembly should proceed. For this use three- or four-lead flexible cable strips (available at most radio stores and some hardware stores). The four-color coding can be supplemented by marking one side of these cables with modeling colors, permitting an identification of each cable, leaving the four colors to identify individual leads within the cable.

In preparing board mounting, be sure to leave ample clearance (a half-inch or so) between the panel and the display board. The minimum spacing between boards is about an inch. Board mounting to the case can be by spacers and screws, but studs mounted by screws through the front panel are probably less troublesome.

INITIAL TEST AND CHECK

The best way to conduct the initial check is in steps, by functional sections. A good place to start is the crystal oscillator. If there is no oscillation, try changing to a different crystal, even if it is not the correct frequency. Crystals will oscillate on their fundamental frequency in this circuit. Oscillation with any replacement crystal indicates a sluggish original 1 MHz crystal. Before discarding a crystal, try changing the value of R3 and R2; this sometimes helps with low-activity crystals. Circuit problems will likely involve the integrated circuit or possibly a wiring error.

When the crystal circuit is working, feed its output to the timebase chain. Oscilloscope tracing should show the successive stages of countdown; a VTVM can also be used to show proper operation. It will read about 300 mV if the stage output is locked to LOW, about 4.5V if it is locked to HIGH. An intermediate value will indicate proper counting. The intermediate value is about 2.5V for the 1, 2, and 4 levels of the counter, and slightly less for the 8 level.

With the timebase working, set the switches to RUN and DISPLAY, and jumper the 1 MHz output terminal to the input. It should now be possible to follow the 1 MHz signal through the input amplifier, the trigger gate, and into the count accumulator. The diodes for the lowest decades should be dimly illuminated. The higher decades should visibly blink. The blinking rate can be changed by shifting the input to the 1 kHz or the 1 Hz point, which allows a visual check of all stages of the counting chain.

When the counter is working, set the control switches to F (frequency), HZ, and DISPLAY: the RUN switch should be in the automatic position. Each 10 seconds, the display should disappear for one second, then appear for one second with blinking, then extinquish, except that the *units* diode of the right digit may remain illuminated for a full eight seconds.

Remove the jumper and feed a few volts at 60 Hz to the input. With the selectors set to P (period), US (microseconds), and DISPLAY, a reading close to 16667 should appear; the exact value will depend on the crystal frequency. On switching to MS (milliseconds) the display should indicate 16 or 17. With the selectors set to F and HZ, the display should indicate 60 (±1). On KHZ (kilohertz) it will probably indicate 1.

At this point it is desirable to make a short check of stability. With a 60 Hz sine wave, a variation of 10 μsec or so in successive readings should be expected. The variation is due to inexact triggering, as established by the sloping sine wave. Changing to a good 60 Hz square wave should decrease the variation between successive readings to about 1 μsec. If a larger variation is found, there is probably a problem with the crystal oscillator, or with stray AC pickup in the timebase circuits. Having a counter with resolution down to 1 μsec or to 1 Hz is a quick way to learn about the problems of oscillator stability.

A preliminary adjustment of crystal frequency can be made at this time. Set the switches to P and US. Adjust C1 until the display reads 16666 or 16667. It may be necessary to place a small parallel capacitor (padder) across C1 to obtain this. If oscillation stops before the correct period is obtained, try adjusting the values of R1 and R2 to restart oscillation.

When these checks have been completed, make the final assembly, then repeat the checks to insure that all wiring is correct. For final adjustment the unit should be allowed to warm up for at least a half-hour. During this time the case will likely become appreciably warm—this is due to the relatively high dissipation in the TTL integrated circuits. The use of an AT-cut crystal minimizes the effect of this temperature change on accuracy.

CALIBRATION

There are two basic methods of calibration. One is to beat the 1 MHz oscillator against a reference frequency, which can be the 5 or 10 MHz WWV signal. (This requires an auxiliary receiver.) Adjusting the series trimmer capacitor of the crystal (C1) should permit attaining an error of no greater than 1 Hz at 1 MHz.

The second method of calibration is to measure a known frequency, adjusting C1 until the indicated value agrees with the known value. A possible source of this frequency is a local broadcast station—the frequency must be accurate to 50 Hz, and is usually within 5 Hz. A call to the station engineer will give the monitor reading, with a probable error of 1 Hz. It may be possible to pick up a local station by connecting 20 or 30 feet of wire to the counter input through a series capacitor of 100 pF or so. Better results will be obtained if a tuned circuit is used.

This adjustment of crystal frequency is the only calibration step required. This simplicity is one of the reasons digital techniques are so popular; since they count, many of the internal operations can be made self-calibrating. In common with all of these digital instruments, an error in count is an indication of some internal trouble. As long as the crystal is on the correct frequency, and things are working properly, the count will be correct.

USE OF THE INSTRUMENT

The obvious use of the frequency meter is for the measurement of any unknown frequency or period. For these, it is only necessary to connect the input signal, making sure that it is over a volt in amplitude (and preferably not more than 10V or so, to avoid overload of the input protectors). The instrument is especially designed for repeated measurement of a continuous signal, through the auto reset feature. Auto reset also works on period measurement, although the time for repetition of the measurement may become quite long.

Many times it is possible to make a measurement as either a frequency or a period. Below 1 kHz measurement of *period* will give greater resolution; above 1 kHz measurement of *frequency* gives the higher resolution. However, it should be remembered that accuracy may not be improved; the true value for either frequency or period is the indicated value, plus or minus error in the clock, plus or minus jitter due to waveshape, and plus zero or one count. With the switches set for F and KHZ, the last term can amount to 1 kHz error.

The procedure to measure a single period—say, the length of a phototube output pulse when a flashbulb is fired—is the following. First, with no input, set the selector to P. For a flashbulb test, the period would be measured in microseconds. Set the selector to RUN, and let the counter cycle once to clear the display section, then return it to normal. After the pulse has been recorded, its duration may be read by switching to DISPLAY. This reading will be held until the unit is recycled or until the power is turned off.

To measure the number of pulses in a train, set the counter to F and RUN, and recycle as above. After the train has ended, turn switch to DISPLAY and read the number of pulses in the train, up to a maximum of 99,999. If desired, up to eight successive periods or pulse trains can be recorded on top of each other. This allows an average to be computed, by division of the *indicated reading* by the *number of periods* (or number of pulse train ensembles) which have been allowed. This pulse averaging could be made automatic by rearrangement of the switching, to include the clock divider in the gate control circuit.

To measure the length of duration of an RF pulse, an envelope detector must be used to convert the pulse into a time period. This envelope detector could be the RF probe of an oscilloscope or VTVM. Remember that an external amplifier may be required if the pulse train amplitude is low, since the counter requires at least 1V for proper operation.

To measure the frequency of a pulse train directly, it is necessary to use a gate period which is shorter than the length of the train. There is also the problem of missing the train completely, or of "chopping it in pieces"; the probability of missing is 90% with the simplified techniques used in this counter. However, several trials will usually give a reading.

Some of these problems can be avoided by using a tracking filter with the pulse train. The audio oscillator for audio and low supersonic frequencies, or the function generator for the very low and audio frequencies can serve for this. It is good practice to observe the output of the filter on an oscilloscope, to determine when the filter is correctly tuned, then to measure the filter frequency using the frequency meter. Incorrect tuning of the filter is indicated by a sudden shift in tracking-filter output when the pulse train occurs. Similar techniques can be used for RF; the color burst oscillator of a television receiver is a good example of the technique needed, and can provide a set of design values for an RF tracking filter.

If the frequency meter is to be used to set an oscillator precisely to some desired value, the 10-second delay between readings will probably be found annoying. One solution for this is to provide an additional position on the *range* switch. This position can be connected to the 10 pps point of the counter chain, to give resolution to 10 Hz and a reading once each second; it could also be connected to the 100 pps point, to give a resolution of 0.1 kHz, with 10 readings per second.

While these techniques of measuring cover most of the common possibilities of measurement frequency or period, others do exist. Experimentation will show many. A review of the extensive literature on counters and on accuracy in frequency measurements is recommended.

FREQUENCY/PERIOD METER SPECIFICATIONS

- Measures frequency from 1 Hz to 15 MHz
- Measures period from 1 μsec to 99.9 sec
- Accurate to 1 part in 10^6 ($\pm$ 1 count)
- Serves as a frequency standard
- Serves as a counter (1 to 99,999 events)
- Automatic recycling

PARTS LIST FOR FREQUENCY METER

C1	—	3–15 pF air trimmer
C2	—	0.01 μF capacitor
C3	—	220 pF capacitor
C4	—	0.01 μF capacitor
C5	—	10 μF, 15V electrolytic
C6	—	0.01 μF capacitor
R1	—	1800Ω All resistors ½W
R2	—	1800Ω
R3	—	470Ω
R4	—	15K
R5	—	1000Ω
R6	—	10K
R7	—	2.2K
R8	—	2.2K
R9	—	1000Ω

R10 — 470Ω
R11 — 1M
R12 — 1000Ω
XTAL — 1 MHz (fundamental)
AT-cut 0 temp coeff

Semiconductors
1 — 1A silicon diode
20 — MV-50 LED devices
1 — 2N4220 FET
1 — 2N706, etc.
2 — 2N3393, etc.
1 — 2N5296, etc.
12 — 7490 decade counter IC
1 — 7404 hex inverter IC
1 — 7400 quad 2-input gate IC
1 — 7413 dual Schmitt AND

Chapter 12

The Sound Level Meter

The sound level meter is an instrument which is not as popular as it deserves to be. Any worker in sound should have one—even the serious hi-fi listener will find it useful. Some of the particular areas of application are in measuring the relative response of speakers and microphones, and in determining speaker patterns and the contours of equal sensitivity of microphones.

Measurement of the noise level of industrial machines is a large and important area which needs to be expanded. Establishment of sound level is an important factor in many psychological tests. The sound level meter is even useful in settling arguments with neighbors when their hi-fi "rocks the air." Equally, it is helpful in setting levels of your own gear to the point where neighbors don't object. More practically, it is a very useful adjunct in balancing the output level of speakers, in measurement of speaker placements in rooms, in professional auditorium work, and in amateur theatrics.

BASIC DESIGN FACTORS

The sound level meter, Fig. 12-1, is intended to provide an absolute measure of the level of the sound impinging on it, over the range of 40 dB to 113 dB above the subthreshold or "zero" sound level of 10^{-16} watts per square centimeter (W/cm^2). Indication is by meter, having a scale range of -20 dB to $+3$ dB, supplemented by a multiplier having steps of 10 dB and a range of 60 to 110 dB; the final reading is the sum of the two. Measurements may be made with a response curve which is flat from very low frequencies to approximately 10 kHz, or may be made with two types of response *weighting*. The

Fig. 12-1. The sound level meter. This unit is smaller than commercial counterparts. Panel is finished in black lettering on white plastic, protected by a clean plastic outer layer.

instrument is self-contained, operating from internal 9V batteries. The construction is similar to others of the miniature lab series, but the instrument is smaller—even smaller than commercial sound level meters. The overall size is 1¾ × 2½ × 5¾ inches, and weight of the instrument with integral batteries is one-half pound.

CIRCUIT AND THEORY OF OPERATION

The basic measurement of sound level requires three elements: a pickup microphone, an amplifier, and an indicator. Because sound levels can vary over an enormously wide range, a practical meter also needs level-setting attenuators. Further, because the human ear does not hear all frequencies equally well, a good level meter needs some means of shaping the frequency response, a process commonly called weighting. (The term comes from precision measurements and really means assignment of an "importance" to a particular measurement or, in this case, a particular frequency.)

The complete level meter circuit, in block diagram form, is shown in Fig. 12-2. The first element is a crystal microphone. This is followed by a weighting network which includes a matching preamplifier source follower, used to place minimum load on the microphone and to give a low-impedance drive to the weighting network. A fixed-gain

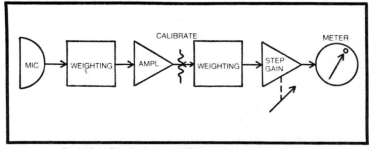

Fig. 12-1. Block diagram of typical sound level meter.

amplifier follows this: feedback is used to establish the upper frequency response of the entire sound level meter. This amplifier is followed by a calibrating gain control, then by a second weighting network. The next element is a variable-gain amplifier, adjustable in steps, with a possible gain variation of 50 dB. The total gain of the two amplifiers is set to compensate for the sensitivity of the microphone used. The output of the final amplifier is AC coupled to a rectifier-type indicator of the fast-charge, slow-discharge type.

The complete circuit diagram is shown in Fig. 12-3. The input impedance for the impedance-matching FET source follower is established by a shunt resistor. A small bypass capacitor is connected across this to reduce the possibility of errors in reading due to RF pickup. The output circuits of the source follower provide a low-impedance driving source for the first weighting networks.

The fixed-gain unit is a 741 operational amplifier, with feedback. In the prototype the gain of this was set to 100, or 40 dB, but this could be increased to as much as $55-60$ dB to compensate for microphone sensitivity. The capacitor across the feedback resistor converts the circuit to an integrator at high frequencies, and provides the upper frequency rolloff desired. The input circuit is switchable to provide two time constants, one with a turning frequency of 125 Hz, the second with a turning frequency of 10 Hz, for nominally flat response. The complete weighting functions are described later.

The second amplifier circuit is also a 741 operational amplifier with gain set by feedback. However, the gain of this amplifier is adjustable in steps. The values are selected to give a gain change of $\times 10$ per step, or 20 dB. In the prototype, the range of gain adjustment is from unity to 3.16×10^2, a total range of 50 dB. The input circuit for this amplifier has three time constants, with turning frequencies of 800 Hz or 225 Hz for weighted response, or of 10 Hz for nominally flat response.

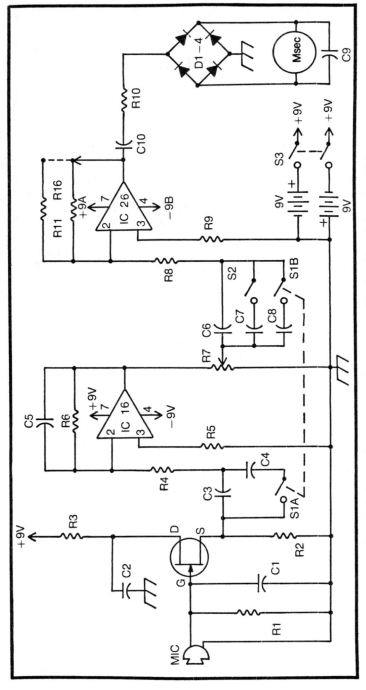

Fig. 12-3. Schematic diagram of the prototype. The functions follow Fig. 12-2.

Resistors R11–R16, used for setting the step gain, should be selected to be within a percent or two of indicated value if good measurement accuracy is desired. If 20% resistors are used, there can be considerable variation in reading from one scale to the next. While this will affect the absolute accuracy of the instrument, it will be useful for *relative* measurements.

The metering circuit at the output of the second amplifier is a bridge rectifier that uses *germanium* diodes. The rectifier output feeds a capacitor which charges rapidly because of the relatively low driving resistance. The discharge rate of the capacitor is set by the size of the capacitor and the resistance of the meter. If desired, two values of capacitance could be provided to give the two time constants recommended for steady-state and "impulse" sounds.

The usage of the weighting networks is dictated by the fact that the sensitivity of the human ear to different frequencies changes with sound level. The weighting networks are intended to approximate this change. One weight is the flat response curve and is intended for use with loud sounds, at 100–110 dB level or higher. The ear hears such sounds with nearly the same sensitivity at all frequencies. The second weight has a single inflection, around 200 Hz, and is intended for medium-level sounds around 60 dB in level. The last weight has two points of inflection and is intended for very low levels of sound; for these the ear has its maximum sensitivity around 1000 Hz.

The specific recommendations of the American National Standards Institute (ANSI) regarding these networks are shown in Fig. 12-4. The recommendations are in the form of upper- and lower-limit curves at each frequency, shown in broken-lines. For this design, these are approximated by straight-line curve segments. The intersections establish the turning points (the 3 dB points of the response curve).

COMPONENTS AND CRITICAL AREAS

Two components of the sound level meter—the meter and microphone—determine the requirements for the gain, case configuration, and mounting detail. The meter used in the prototype is a 0–200 μA meter, which edge-mounts in a nominal ½ × 1½ inch rectangular hole. It was originally designed as a tape recorder level meter, and was obtained from a surplus supply house. Its major advantage is that the dial is precalibrated in the desired logarithmic scale. Identical meters are advertised from time to time. These should also be available as a spare part. Similar 0–200 μA and 0–1 mA

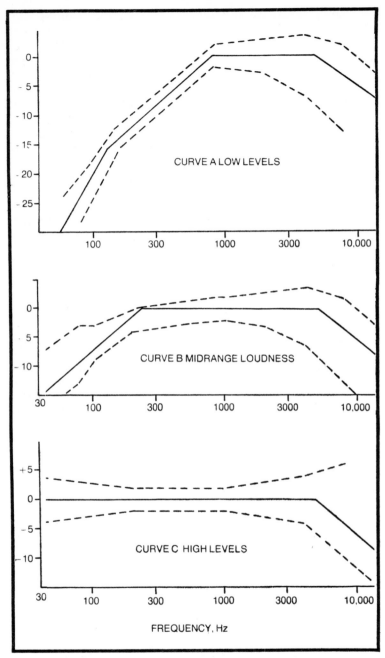

Fig. 12-4. ANSI recommendations for noise weighting limits (broken lines) and design values used for the prototype (solid).

meters are also available from several sources, but without the special VU scale markings. There are other types of meters with scale markings in decibels available in surplus from time to time. The exact meter characteristic is not of great importance. Any sensitivity from approximately 50 μA to 2 mA full scale is usable. The calibration range of the amplifiers is adequate to compensate for the variation. The only really critical requirement for the meter is that it should not be highly damped; electrical damping is used to provide the damping characteristics desired. Meters intended for level indication meet this requirement.

The most important component is the microphone. This should be a diaphragm-type ceramic, crystal, or condenser microphone. The unit in the prototype is contained within a small plastic case. This unit is also a surplus component, purchased as a kit of four identical units. Identical microphones are offered by several surplus parts houses, in identical size case, sometimes with cases of different color. Several catalogs list similar types.

The parts list gives several sources where these two components might be obtained. It is recommended that these components be obtained first, and the remainder of the instrument built around them. If identical size components are found, then the case dimensions and layout can be followed exactly; if not, some adjustment will be needed.

Because of the small amount of space available, all of the large capacitors should be of the tantalum type. These are available at a reasonable price from most local distributors and nearly all of the large mail-order supply houses.

If the meter to be used has a linear scale, it will be necessary to modify it to reflect decibel increments. Table 12-1

Table 12-1. Application of Decibel Increments to Linear-Scale Meter.

LEVEL, dB	METER INDICATION, % OF FULL SCALE
−00	0
−20	7.07
−15	12.6
−10	22.4
−5	39.9
−3	50.1
−1	64.1
0	70.7
+1	79.9
+2	89.2
+3	100

gives the relation between indicated decibels and percentage of full-scale reading. If the meter case can be opened, the decibel or VU scale can be marked on the dial face; as an alternative a new face can be made up. If the meter is hermetically sealed, the scale can be placed on the panel—for example, above the top of a horizontal (edge-mounted) meter.

GENERAL CONSTRUCTION

The general construction of the sound level meter is shown in Fig. 12-1. The case is smaller than most instruments of the miniature lab—a minibox of $1\frac{3}{4} \times 2\frac{1}{4} \times 4$ inches. Incidentally, the case used in the prototype illustrates one of the traps awaiting the unwary: The board layout was made to standard case dimensions, taken from a table. When assembly time came, I found that the case from the particular supplier was nearly $\frac{1}{8}$ inch smaller in width than the nominal standard. Rather than scrap the case, I bent up a new lower half on a vise to fit the outside of the front panel section.

Figure 12-5 shows the construction of the microphone mount. The components of this are made from standard PVC fittings (available at hardware stores). The bottom section is a molding, a tube-to-threaded-pipe adapter. The threaded section of this is cut off and the remainder mounted on the aluminum case by three self-tapping screws. (A pipe cap would be equally good.) The next section is a length of pipe cemented into the adapter. The mike is held in this one-inch pipe section by a setscrew (not visible in the photographs). The offset is intentional, to allow a calibrator section to be slid over the microphone, giving a specific reference point for spacings.

As obtained, plastic pipe fittings are sometimes stained or marked in someway with indelible ink. Steel wool removes the

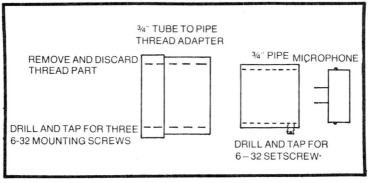

Fig. 12-5. Microphone mounting as used in the prototype.

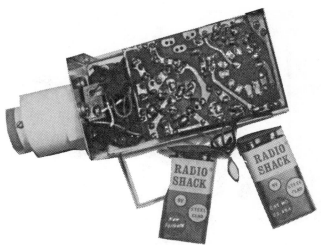

Fig. 12-6. Internal view of prototype with outer case removed.

stains and markings and leaves a nice finish which does not require painting.

The internal construction of the sound level meter is evident from Figs. 12-6 and 12-7. The PC board is shorter than the case, allowing space for the meter and its mounting. A single clip holds the meter in the prototype, but screw mounting could have been used equally well. The PC board is supported by the *sensitivity* switch, and in turn from the front panel. Batteries are held between the case back and the PC board by an insulating block of rubber foam.

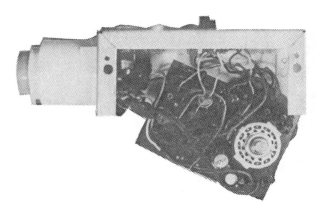

Fig. 12-7. Disassembled view of prototype. The circuit board is supported by the level selector switch and braced by the foam rubber sheet used between batteries and board.

191

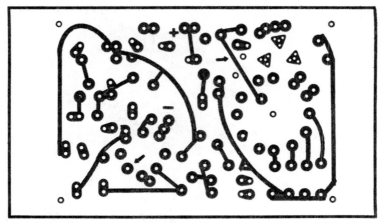

Fig. 12-8. Full-size printed circuit board layout, prepared with "Easy-Etch" transfers. (Compare with board photo of Fig. 12-6.)

Circuit board layout is shown in Fig. 12-8, and parts location on this board in Fig. 12-9. Resistors and other small components are mounted vertically, being lead-supported, as is common in subminiature radios. There is no space problem if tantalum capacitors are used as recommended. Aluminum electrolytics could be used but would require some squeezing of components. Leads and jumpers are shown in Fig. 12-10.

The rotary switch is a single-pole 12-position switch, with terminals parallel to the shaft. The terminals should be

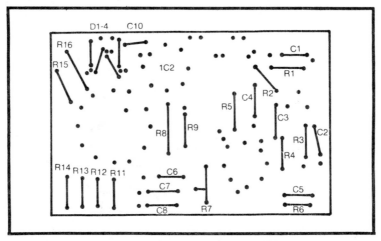

Fig. 12-9. Drilling and component mounting arrangement. Unused holes allow for additional ranges and for an offset-null potentiometer, if needed. (Compare with component-side board photo of Fig. 12-7.)

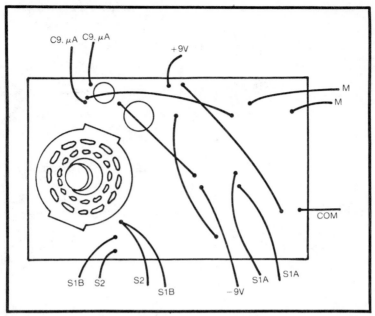

Fig. 12-10. Jumpers and lead arrangement. The lead marked COM is common to the two batteries.

prepared as shown in Fig. 2-9, for conversion to printed circuit board use.

The panel layout is evident from Fig. 12-1 and from the drawing of Fig. 12-11; the meter opening indicated should be changed to suit the particular meter available. The original unit and the drawings show only three switches, since the prototype provided only a single meter damping. If two values of damping are to be provided, the panel layout should be modified to place four slide switches in a row instead of the three shown. There is ample room for this addition. The panel in the prototype is finished in the plastic-on-plastic technique described in Chapter 2. Black lettering on a white background is used, but could be any other color.

CHECKING AND ADJUSTMENT

The easiest way to check out this unit is by sections. Initial checking should be done with the board assembly out of the case, starting without the microphone. Set *calibration* control R7 (board-mounted trimmer) to zero and feed a signal of about 1 mV at 1 kHz to the microphone leads. Signal should appear at the output of the first amplifier, at a level of approximately 100 mV. The exact value is not important, but an appreciable

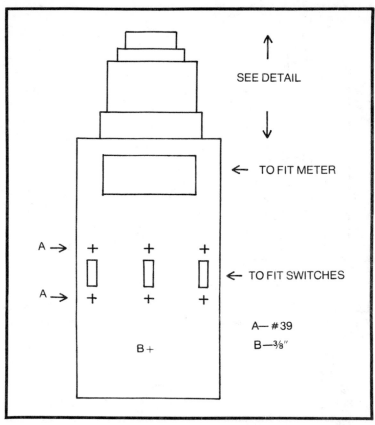

SEE DETAIL

← TO FIT METER

A →
A →

← TO FIT SWITCHES

A— #39
B—3/8"

B +

Fig. 12-11. Panel layout. The meter hole should be adjusted in size and location as needed.

difference probably indicates trouble. Next, set the level switch for minimum attenuation (60 dB). Advance the *calibration* control; meter deflection should start almost immediately. Set the input level to 0.36 mV, and adjust the calibration control for a deflection of 0 dB on the meter (70% of full scale if a linear meter is used). This step calibrates the electronics for a microphone having a sensitivity of −55 dBm, an average microphone.

Measure and record the input required for each of the major meter markings. Due to nonlinearity in the rectifiers, it is likely that the low end of the meter scale will be in error; for example, the −20 dB point may occur at 12% of full-scale input rather than the ideal 7%. Above about midpoint on the instrument the readings should be close to the ideal values given in Table 12-1.

Check the error from scale to scale by changing the input by a factor of 10 dB (×3.16 in voltage) for each scale step. If the resistors have been selected to within 1% of the design value, the error between scales should be of the same order. Check also the response curve at different frequencies, for the three weight settings. The response curves should be close to the design curves of Fig. 12-4. If the response does not fall between the limit curves, it is probable that the capacitor values are incorrect. However, for correction it may be easier to change the associated resistors.

Damping of the meter should cause it to reach 99% of full scale in 0.3 second with overshoot of 1−1.5%. This can be checked by setting the meter to full scale with an input signal at 400 Hz fed through the tone-burst generator. The change between the 128 cycle *on* and the *continuous* position should be small, but the meter reading should decrease as the number of cycles in the burst decreases below 128.

For a final check, connect the microphone and set the attenuator to 70 dB. If the microphone is close to average sensitivity, normal conversation should cause the meter to swing to approximately the 0 dB point, with the meter held about 3 feet from the sound source. The value will vary with the talker, and with microphone sensitivity. Of the microphones checked in the prototype, all coming from a single kit, there was less than 6 dB variation.

With everything working, complete the final assembly, then recheck by repeating the voice test. The unit is now ready for final calibration.

CALIBRATION

The following describes different ways of calibrating the instrument, ranging from an approximate calibration to an absolute calibration capable of high accuracy. Except for the simplest methods, the description is introductory.

The simplest method of all is to use the approximate calibration already established. The error in this is the deviation of microphone sensitivity from that assumed in the calibration. If the meter is to be used only for relative measurements, such as balance of speaker outputs, the error is unimportant.

The second method of calibration approximately compensates for the microphone sensitivity as well as other factors; it involves checking a number of noise sources. Table 12-2 shows typical sound levels. An approximate calibration can be prepared by adjusting the meter sensitivity to give these readings, on the average. For example, the noise level

Table 12-2. Approximate Sound Levels of Common Sources.

SOUND	AT DISTANCE, ft	LEVEL, dB
Ordinary conversation	3	65
Average automobile	15	70
Milling machine*	25*	75
Motor truck	15	80
Street traffic (busy hour)	15	85
Jackhammer	10	95
Jet Plane (large)	50	100
*ASSUMING INDOOR MEASUREMENT, IN ROOM OF 50 × 100 ft.		

produced at a distance of 15 feet at city traffic speeds should be close to 70 dB, so the noise level meter is adjusted to give this reading, on the average. Conversation can also be used for calibration, taking the average of several persons. However, it is difficult to avoid self-consciousness, which affects speech patterns—a problem which does not exist with mechanical noise sources. This method of calibration is satisfactory for all casual use, and for all relative measurements.

The third method of calibration is by comparision with another sound level meter. Schools and industrial shops often have such meters. A few hi-fi stores have units for their own use, and many of the national chain stores have commercial versions for sale. To make a comparison, the meters should be placed side by side, and at least 5 feet away from a noise or sound source. Tones give the most accurate calibration. With this relative calibration, it should be possible to bring the error to within a few decibels.

A variation of this is to use a *calibrator*—an oscillator/amplifier/loudspeaker having known output. These transfer standards ere often found in the acoustical laboratory of schools and industry. Sometimes it is possible to arrange for their use on a temporary basis.

The final method of calibration is the most accurate: it compensates for all factors in the instrument and gives an absolute calibration. This method depends on the fact that many types of microphones are bilateral devices—that is, they are equally good as microphones or loudspeakers. The class of microphones with which this method can be used are called *reciprocal transducers*, and the method is called reciprocal calibration. Crystal microphones belong to this class, which is one reason they are specified for the sound level meter.

The complete instrument calibration requires two steps: calibration of the microphone by the reciprocal method, then the setting of the amplifier gains to the proper value.

Reciprocal calibration requires three microphones and the ability to place these in known positions. The microphones do not need to be identical, but uniform physical size is convenient, for the simple mechanical reason of mounting. The specific steps of measurement are shown in Fig. 12-12, and the relations involved in Fig. 12-13. Two electrical quantities are measured: the output voltage of the transducers used as microphones and the excitation current of the transducers

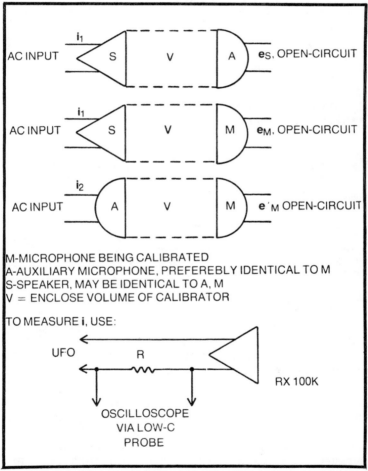

Fig. 12-12. Principles of reciprocity calibration. The measurements on the three pairs of microphones provide the data for absolute sensitivity calibration.

Calculation of Microphone Sensitivity:

Density of air at 20°C, $\rho = 0.0012$ gm/cm^3
Velocity of sound, $\quad \nu = 331.4 + 60.7\,\mathrm{T}\nu$ m/sec
$\qquad\qquad\qquad = 344$ m/sec at 20° C

1. *Calculate Acoustic Capacitance of Coupler of Fig. 12-14:*

$$C_A = \frac{V}{\rho C^2} \quad (V = \text{Coupler Volume, Cm}^3)$$

2. *Calculate Microphone Sensitivity from Measurements of Fig. 12-12:*

$$k_m = 3.16 \times 10^{-3} \sqrt{\frac{2\pi f \cdot C_a \cdot e_m \cdot e'_m}{e_s \cdot i}} \qquad \frac{\text{Volts}}{\text{dyne-cm}^{-2}}$$

Reference: Olsen, Acoustics

: Tremaine, Audio Cyclopedia

Fig. 12-13. Relationships for using the measured data to establish the reciprocal calibration.

used as loudspeakers. The acoustic coupling between the units is determined from the volume of the coupling cavity combined with air density and the velocity of sound. These last quantities are determined by the barometric reading and the temperature, and needed only for best precision, since the normal atmosphere density and sound velocity vary only by a few percent.

When the microphone sensitivity is determined, the output for a given signal level is calculated. For example, if the sensitivity is −60 dB, the output would be 1 mV for a 74 dB sound level, or 10 mV for a +94 dB level. The amplifier gain is adjusted to give these values, i.e., to read 74 dB with a 1 mV input (flat response weighting).

Figure 12-14 shows the construction of a calibration adapter. The basic element is a plastic pipe coupling. The other elements are mountings for two microphones, similar to the mounting on the sound level meter, and are intended to hold other calibration transducers. If the microphones are not identical, the mounts should be tailored to fit.

Well equipped laboratories will have a commercial absolute microphone calibrator—for example, the General Radio 559-B unit. Smaller laboratories may have a relative calibrator, such as the Hewlett-Packard 15117A and the General Radio 1562A. It may be possible to arrange for the use of one of these special instruments.

For casual use of the sound level meter, the first two calibration methods should be used; for serious use, the *absolute* method should be used, either with a home-built coupler or with one from a commercial laboratory. Regular users might wish to make up a relative calibrator, and use this as a transfer standard. A simple single-frequency oscillator, plus the assembly of Fig. 12-14 is adequate.

USES OF THE SOUND LEVEL METER

There is one particular use of the sound level meter that I would like to see become more common—this is in the

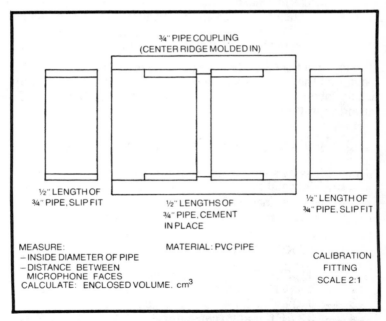

Fig. 12-14. Adapter to couple two microphones, having a known common volume. May be used for reciprocity calibration or with a loudspeaker-/oscillator combination used as a relative level calibrator.

purchase of mechanical equipment. Measurement of the noise produced by air conditioners, dehumidifiers, humidifiers, electrostatic air cleaners, electric fans, dishwashers, vacuum sweepers, automobiles, and other mechanical devices can prevent buying a source of annoyance. If just a few purchasers started making these measurements, manufacturers would take more pains with noise quieting than is now taken.

An associated sound level measurement is taken during installation. For example, the meter will indicate when an external air conditioner will be an annoyance to building inhabitants and to inhabitants of neighboring buildings. There is no real justification for the existing level of noise pollution.

Some of the additional measurements which can be made are:

- Sound dispersion patterns of speakers
- Uniformity of sound distribution in auditoriums
- Acoustic decay time in auditoriums
- Many other problems of public address installations

For the homeowner, some of the uses are:

- Making comparisons of speakers or microphones before purchase
- Establishing location of speakers for good sound reproduction
- Setting the balance of stereo speakers

The use mentioned in the introduction, of setting levels to the point that they do not cause annoyance to neighbors, was only partly facetious—this also is a worthwhile use of the meter as an aid in preventing noise pollution.

SOUND LEVEL METER SPECIFICATIONS

- Measure sound levels from 40 to 113 dB
- ANSI weighting, A, B, or flat response
- Self-contained with integral microphone
- Designed for *absolute* or *relative* calibration

PARTS LIST FOR SOUND LEVEL METER

C1 — 30 pF ceramic disc capacitor
C2 — 4 μF, 15V electrolytic
C3 — 0.33 μF Mylar or tantalum capacitor
C4 — 3.3 μF tantalum capacitor
C5 — 200 pF ceramic disc capacitor
C6 — 0.1 μF Mylar capacitor

C7 — 0.35 μF Mylar or tantalum capacitor
C8 — 3.3 μF tantalum capacitor, 35V
C9 — 100 μF, 15V electrotytic
C10 — 10 μF, 15V electrolytic
R1 — 1M   All resistors ½W
R2 — 2.2K
R3 — 100Ω
R4 — 10K
R5 — 10K
R6 — 1M
R7 — 500 PC trimmer potentiometer
R8 — 10K
R9 — 10K
R10 — 6.8K
R11 — 10K
R12 — 22K
R13 — 68K
R14 — 220K
R15 — 680K
R16 — 2.2M
M1 — Crystal condenser, or ceramic microphone
D1-4 — Germanium small-signal diodes
IC1,2— 741
μA — microammeter (any value from 100 μA to 1 mA)

Chapter 13

Useful Bits and Pieces

Around the lab there is always need for small items, not properly called instruments, although some are calibrated, and not properly called tools, although some are useful in servicing and repair. Let us call these *bits and pieces* and use the following as examples of items which are worth constructing.

LOGIC PROBE

In troubleshooting logic or digital circuits, there is often need to tell the state of a gate, whether it is *high* or *low*. Of course, one can drag up an oscilloscope, or even use the lab VTVM. These work, but take time, the time needed to find the right pin with the probe, of looking over at the scope or meter face, and then back for the next pin. During this time there is a fair chance that the probe tip will slip off the right pin, possibly causing damage. What is needed is a probe with the indicator built into it—a *logic probe*.

There are some beautiful examples of these probes on the market. Hewlett-Packard, for example, has one which slips over a 16-pin DIP, and which shows the state of each of these terminals by the glow of an LED. Several companies have quasi-computing probes, which can be set to match the IC type being tested. But for occasional use, these are much, much too expensive; what is needed is a simple pencil-type logic tester—they are available but still at relatively high cost. Here is a minimum-cost version, as shown in Fig. 13-1.

The probe is designed to work only with TTL (transistor—transistor logic) units, and is based on the fact that the output of these can source or sink considerable

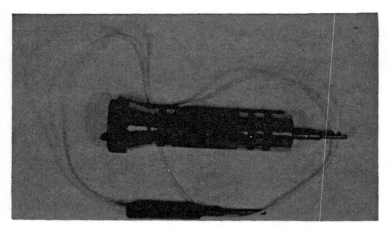

Fig. 13-1. Tᴛʟ logic probe. The lead goes to common, the probe tip to the gate terminal. Lᴇᴅs in the cap indicate high, low, or both.

current. The probe itself is very simple—basically a battery and two LEDs, plus a protective resistor and a case. The circuit is shown in Fig. 13-2. The two LEDs are connected back to back, in series with the resistor and the 1.5V battery.

In use, the common lead goes to the negative or common bus. When the probe tip is also placed on this bus, the battery discharges through one LED, which lights up. If the probe is on a logic *low*, about 300 mV, current through the diode decreases but there is still enough to light up the same LED. If the probe tip is placed on the +5V bus, there is a net of 3.5V forcing current in the opposite direction, so the second LED lights up, the battery being charged by this current. On logic *high*, at about 3.5V, there is still 2V available to cause reverse current flow, so the second LED lights, but more dimly. If both states

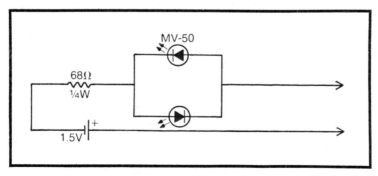

Fig. 13-2. Circuit of the logic probe. One diode emits when the probe is on **high**, the other when on **low**.

are present due to switching, both diodes will light, the brilliancy varying with the duty cycle.

Construction of the probe is as simple as the design. The major parts are shown in Fig. 13-3. The case is a one-cell flashlight—the type using AA cells. The one used for the prototype is from the dimestore (nomenclature from old days). The LEDs and resistor are mounted on a small PC board, which is mounted in the plastic cap of the flashlight, replacing the flashlight bulb. The particular cap has a lip, which protects the diodes and resistor. The lip also provides some shielding from stray light, helpful when illumination level is high. The circuit-board layout and the general arrangement of the flashlight are shown in Fig. 13-4. In the prototype the metal shell holding the spacer and center contact of the battery is the original lamp carrier, with its threads flattened out with roundnose pliers. Note that the battery is mounted upside down with respect to normal usage; with the regular polarity, the case would be the common point and it would be necessary to look away from the point of test to see the status of the diode.

The use of the probe is very simple: clip the lead to a common point and start probing. In doing this, try to avoid shorting two IC contacts together. While most ICs are well protected, shorts may occasionally cause damage.

If LEDs with a reverse voltage specification of 5V or more are available, it is possible to equalize the light output on the *high* and *low* state. To do this, use two resistors, one in series

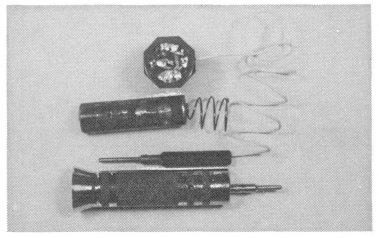

Fig. 13-3. Disassembled probe. The basic flashlight is sold in dimestores. Note that the battery is reversed from its normal position.

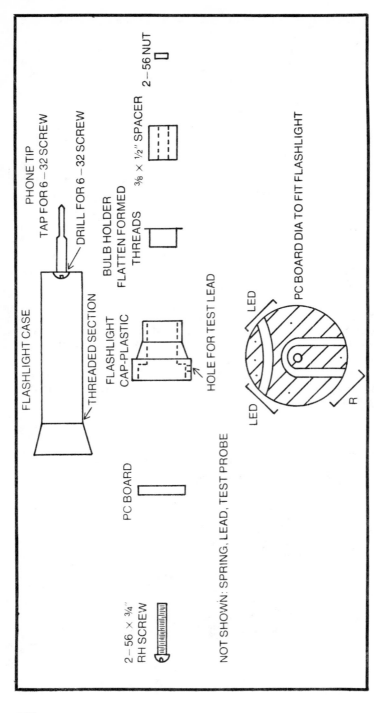

FLASHLIGHT CASE

PHONE TIP
TAP FOR 6 – 32 SCREW

DRILL FOR 6 – 32 SCREW

THREADED SECTION

BULB HOLDER
FLATTEN FORMED
THREADS

3/8 × 1/2" SPACER

2 – 56 NUT

FLASHLIGHT
CAP-PLASTIC

HOLE FOR TEST LEAD

PC BOARD

LED

LED

PC BOARD DIA TO FIT FLASHLIGHT

R

2 – 56 × 3/4"
RH SCREW

NOT SHOWN: SPRING, LEAD, TEST PROBE

Fig. 13-4. Exploded view of cap assembly, case, and circuit board. Some modern phone tips may need to be sweat-soldered to the mounting screw.

with each LED adjusting values for equal brightness. Another way to accomplish this is to use a rechargeable battery with a 2V rating.

The exact appearance of the finished probe will depend on the flashlight available; the type shown here is available generally.

FILTERS

An assortment of filters is always useful around the laboratory. In the audio range, some typical needs are:

- Narrowband 400 Hz and 1000 Hz bandpass filters, to give good waveform for bridge and distortion measurements.
- 150 Hz high-pass filter to eliminate hum due to the 60 Hz line and its second harmonic.
- 2:1 bandwidth or *octave-band* filters for noise analysis and vibration analysis.
- Bandstop filters at 60 Hz, 400 Hz, and 1000 Hz to eliminate specific signals.

Where requirements are not too severe, a three-element *Butterworth* filter is a reasonable choice. The out-of-band attenuation increases at the rate of 18 dB per decade—not exceptionally high, but often useful. Butterworth filters are free of ripples in the passband, so it is only necessary to know the loss at one frequency. With good coils, the loss will be small.

The prototype Butterworth low-pass filter is shown in Fig. 13-5, together with sufficient information to allow you to scale the element values for different impedances and frequencies; remember that frequency is measured in *radians per second*, equal to the frequency in hertz multiplied by 2π. The relationships shown permit determination of values needed for any low-pass filter design operating at any frequency.

Often the need is for a bandpass or a bandstop filter rather than a simple high-pass or low-pass type. The bandpass type is obtained from the low-pass prototype by placing a coil across each capacitor, plus a capacitor in with series with each coil. The relationships for this and some special equations for calculation are also shown in Fig. 13-5.

For bandstop filters, it is simpler to start with a T-section high-pass prototype. The bandstop design is obtained by placing a coil across each capacitor and a capacitor in series with each coil. Figure 13-6 shows the prototype T-section filter and its scaling rules, plus the bandstop configuration and its design equations.

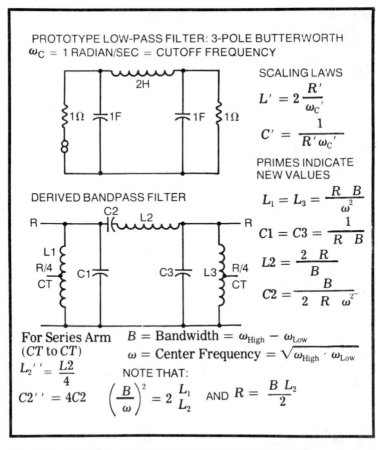

PROTOTYPE LOW-PASS FILTER: 3-POLE BUTTERWORTH
ω_C = 1 RADIAN/SEC = CUTOFF FREQUENCY

SCALING LAWS

$$L' = 2\frac{R'}{\omega_C'}$$

$$C' = \frac{1}{R'\omega_C'}$$

PRIMES INDICATE
NEW VALUES

DERIVED BANDPASS FILTER

$$L_1 = L_3 = \frac{R\ B}{\omega^2}$$

$$C1 = C3 = \frac{1}{R\ B}$$

$$L2 = \frac{2\ R}{B}$$

$$C2 = \frac{B}{2\ R\ \omega^2}$$

For Series Arm B = Bandwidth = $\omega_{High} - \omega_{Low}$
(CT to CT) ω = Center Frequency = $\sqrt{\omega_{High} \cdot \omega_{Low}}$

$$L_2'' = \frac{L2}{4}$$ NOTE THAT:

$$C2'' = 4C2$$ $\left(\frac{B}{\omega}\right)^2 = 2\frac{L_1}{L_2}$ AND $R = \frac{B\ L_2}{2}$

Fig. 13-5. Basic Butterworth low-pass filter and scaling laws; derived bandpass filter and equations for it. The bandwidth ratio of the bandpass filter is determined solely by the ratio of inductances.

For the audio-frequency range, a convenient source of inductors are the toroids used on telephone lines as leading coils, which are readily available surplus items. One of these coils is shown in Chapter 3 (Fig. 3-4). These 44 mH and 88 mH coils are widely available at very reasonable cost. These toroids are very good inductors, having a Q around 150 at a frequency of 8 kHz. The Q decreases approximately by a factor of 3 as the frequency changes by a factor of 10, so a Q of greater than 10 is available over most of the audio range. The coils include two windings which can be used in series or in parallel so that inductance values of 11, 22, 44, and 88 mH are available. The coils are easily modified to secure other values of inductance by removing or adding turns or by connecting coils

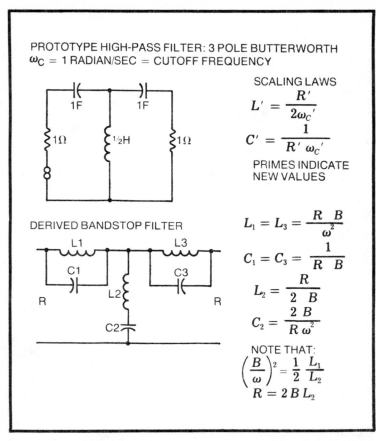

PROTOTYPE HIGH-PASS FILTER: 3 POLE BUTTERWORTH
$\omega_C = 1$ RADIAN/SEC = CUTOFF FREQUENCY

SCALING LAWS

$$L' = \frac{R'}{2\omega_C'}$$

$$C' = \frac{1}{R' \; \omega_C'}$$

PRIMES INDICATE
NEW VALUES

DERIVED BANDSTOP FILTER

$$L_1 = L_3 = \frac{R \; B}{\omega^2}$$

$$C_1 = C_3 = \frac{1}{R \; B}$$

$$L_2 = \frac{R}{2 \; B}$$

$$C_2 = \frac{2 \; B}{R \; \omega^2}$$

NOTE THAT:

$$\left(\frac{B}{\omega}\right)^2 = \frac{1}{2} \frac{L_1}{L_2}$$

$$R = 2 B L_2$$

Fig. 13-6. Basic Butterworth high-pass filter and its scaling laws; the derived bandstop filter and its design equations.

in series. Figure 13-7 gives the approximate inductance versus the number of turns removed or added to the basic 44 and 88 mH toroids.

A convenient way of designing filters which use these coils is to set the inductance to one of the above standard sizes. To do this, it is necessary to let the impedance vary. Capacitance C is chosen to give the desired cutoff frequency and bandwidth. The resulting impedance can be matched to the external circuit by several techniques. One is to make use of the centertap of coils, possible on some filters, which gives a 4:1 impedance stepdown. A second method is to wind an additional coil on the toroid core, using this for input and output. The number of turns is calculated from the relation that the *impedance is proportional to the square of the turns ratio.* A

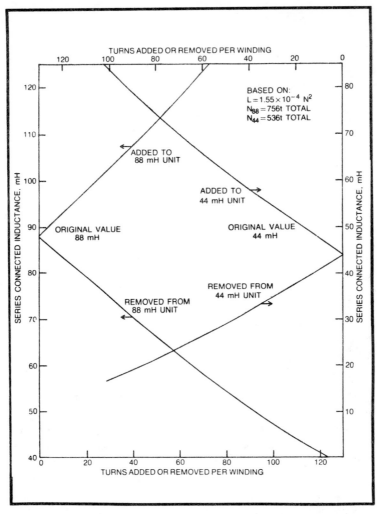

Fig. 13-7. Inductance of 44 and 88 mH telephone toroids versus number of turns added or removed. Example : to secure 100 mH added 25 turns to each winding of an 88 mH toroid (and connect all winding in series aiding).

third way of matching is by use of an external matching transformer.

Which can be used will depend on the filter configuration, and on the range of impedances to be changed. The tapped coil is most useful with bandpass filters operating around the center of the audio range. The added winding is most useful for low input and output impedances, again with bandpass filters.

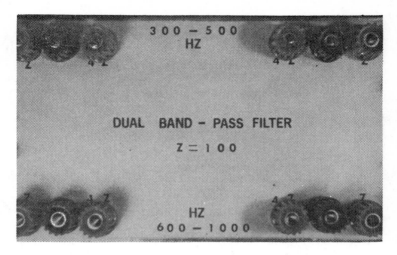

Fig. 13-8. A dual bandpass filter in a single case, with connections for $Z=100\Omega$ and $Z=400\Omega$. (This photo illustrates the need to be careful with placement of labels.) Finish, black lettering on white background.

Transformers for matching can be used with any configuration. In selecting the transformer, note that it is not neessary that the transformer be designed for precisely the impedances to be matched. For example, a 500Ω to 8Ω output transformer will give a perfectly good match between 250Ω and 4Ω lines. At low signal levels, a wide variety of transistor transformers are perfectly satisfactory for this use.

An example of a filter designed to these ideas is shown in Fig. 13-8. This is a dual bandpass filter in a single case. One filter is intended for the input circuit of amplifiers, to give a high-purity sine wave for distortion measurements. The second filter is for the output circuit, to select the second harmonic for distortion measurement. The combination is also useful for intermodulation meaurements—say, by feeding the amplifier with the frequencies of 2000 and 2500 Hz (added together), and measuring the 500 Hz intermodulation present at the output, with the filter selecting this.

The response of these filters is very close to theoretical, as shown in Fig. 13-9. This figure also illustrates a common problem in filter measurements, or in distortion measurements: the error introduced by unwanted harmonics. The low-frequency rejection appears to be very poor when total voltage is measured, even though the oscillator used for signal input is reasonably good. It is only when the input distortion is eliminated that the true attenuation curves of the filter can be seen.

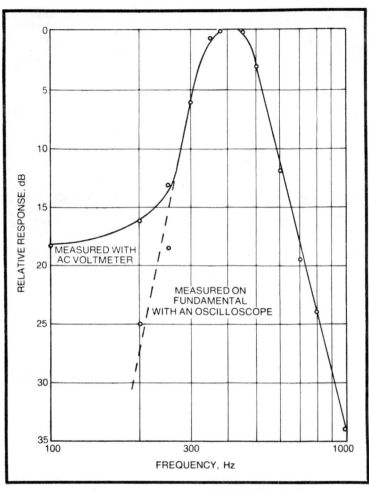

Fig. 13-9. Measured response of one section of filter of Fig. 13-8. Solid line: apparent response, measured with source having approximately 1.2% harmonic distortion (nonselective voltmeter). Dotted line: true response, measured with same source and an oscilloscope used to simulate a tuned voltmeter.

Mounting and general construction of these filters is evident from the photos of Fig. 13-10, with Fig. 13-11 showing a suitable PC board. A set of these filters for octave (2:1) or half-octave ($\sqrt{2:1}$) frequency bands is very useful in noise analysis.

These general principles of filter design may be extended to radio frequencies up to about 50 MHz. The easiest inductors to use in the RF range are toroidal coils, wound on cores with

Table 13-1. Parts List for Bandpass Filter.

PART	RANGE: 300₋500Hz	RANGE: 600₋1000 Hz
C1	2.0 μF	1.0 μF
C2	1.0 μF	0.5 μF
C3	2.0 μF	1.0 μF
L1	88 mH toroid	44 mH toroid
L2	2×88 mH	2×44 mH
L3	88 mH	44 mH

suitable RF characteristics. The various suppliers, *Amidon* and *Indiana General*, for example, supply data for determining the core type, size, and number of turns to use for a given inductance.

The simplified filter designs described here cover many needs, but at times high-performance filters are required. For design of these, see the specialized books on filter design and construction.

PINK-NOISE FILTER

If octave-band filters are used for frequency response measurements using noise as the signal source, the response appears to increase with frequency, even though the system measured is actually flat. This occurs because noise that has uniform intensity (*white noise*) has constant energy per unit of bandwidth and therefore constantly increasing energy for the 2:1 change of octave bandwidth. To secure a constant indication with a flat response system, the noise input must be modified, or weighted, to have constant energy per octave. Such weighted noise is usually called *pink noise*.

Fig. 13-10. Internal view of filter. The thin object between boards is a grounded electrostatic shield, formed from the thin aluminum of a soft drink can and insulated by adhesive plastic film.

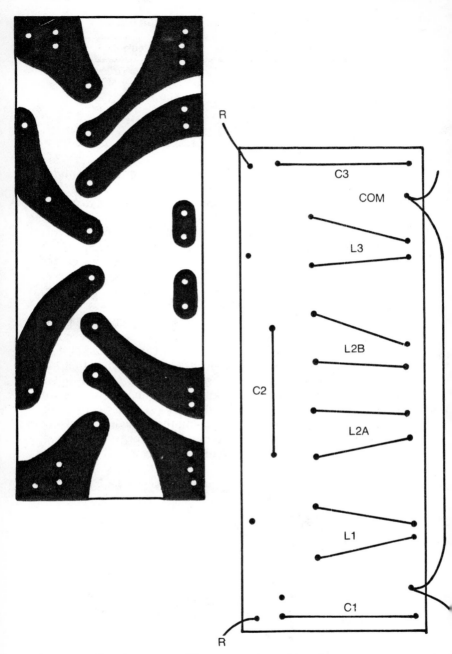

Fig. 13-11. Board layout for filters, component placement, and connections. Pads may be drilled for additional parallel capacitors. Original boards drawn with an ink-resist pen.

A simple circuit to produce this noise is shown in Fig. 13-12. It is derived from the filter used in the General Radio 1390 pink-noise instrument. This filter is conveniently constructed on a circuit board, a suitable layout being shown in Fig. 13-13. This also shows parts locations. The pads are laid out to allow paralleling of capacitors and placing two resistors in series to simplify attaining the correct component values. The board can be mounted in a small case, such as the one used in Fig. 6-7.

In use, the filter must be driven from a relatively low-impedance source, less than 1000Ω. The output circuit should not be loaded, since it is designed only for such high-impedance circuits as electronic voltmeters—the minimum load resistance 20K. The driving source must supply flat noise if the output is to be the correctly weighted pink-noise. A reasonable source, almost always available, is the video-detector test point on a TV receiver tuned to an idle channel. Another possible source is the AM detector output of a hi-fi receiver, with the speaker output being an alternate. The output of an FM detector is not satisfactory, since its noise output increases as frequency increases. Of course, a specialized noise generator may be used if it is available.

One of the areas of application of the pink-noise filter is in the analysis of the noise spectrum produced by machinery. For this, the noise would be divided into octave bands, using filters of the type described above. Other areas of application are the measurement of noise levels on telephone circuits, the noise output of amplifiers, etc. Amplifiers are sometimes tested for ability to carry signals of high peak-to-average power characteristics by using a noise input signal: a change in peak-to-average ratio shows poor performance. Intermodulation measurements are sometimes made by removing the noise over one frequency band with a bandstop

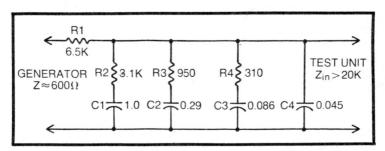

Fig. 13-12. Circuit for converting equal-amplitude white noise to pink noise having constant energy per octave.

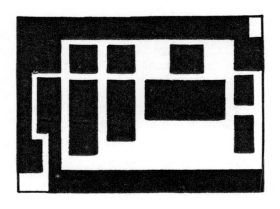

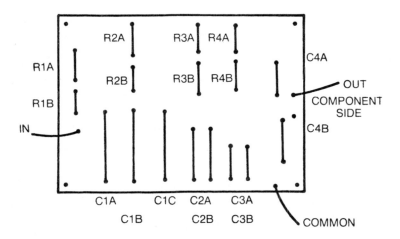

Fig. 13-13. Circuit board for pink-noise filter, component placement, and connections. Pads are laid out to simplify securing exact values by series resistors, parallel capacitors. Drawn with ruling pen and India ink.

filter, and feeding this signal to an amplifier, then measuring the noise level within the band rejected; the noise in this band is now due to cross modulation in the amplifier under test. There are numerous other noise measurements and measurement techniques.

The circuit layout of Fig. 13-13 is also suitable for use with a different type of noise filter, that of Fig. 13-14. As shown in Fig. 13-15, this is intended to approximate the frequency response of the human ear. When measurements of the signal-to-noise ratio of a communication receiver is made, the most reliable indication of this figure of merit is obtained with

216

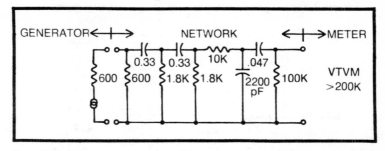

Fig. 13-14. Filter to give noise weighting approximating response of the human ear, per ANSI recommendation. The board of Fig. 13-13 can be used for mounting these components.

a filter of this type interposed between the receiver and the output meter.

CRYSTAL CHECKER/ MARKER/FREQUENCY STANDARD

An oscillator which will operate with any crystal which may be plugged into it is useful as a crystal checker, and also useful for the generation of marker frequencies or as a transfer standard. An easy way to secure this capability is to use a multivibrator circuit, with the crystal replacing one of

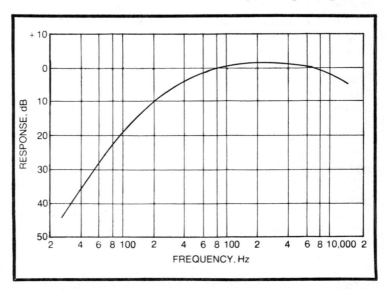

Fig. 13-15. Design response of noise weighting filter. The measured response should be within ±2 dB of this curve, relative to the response at 1000 Hz.

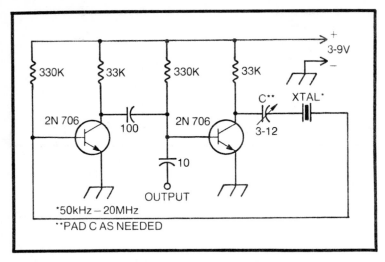

Fig. 13-16. Multivibrator-type crystal oscillator, suitable for a working standard, a crystal tester, or for building into equipment. A crystal which does not oscillate in this circuit is almost certainly defective.

the coupling capacitors. Figure 13-16 shows a suitable circuit. For general use this should have a number of crystal sockets connected in parallel, so that it is useful with many types of crystals.

To use this unit, plug in the crystal and connect a 3—9V battery. The crystal, if good, will oscillate on its fundamental frequency. The circuit has sufficient regeneration so that it may be used with virtually any crystal over the range from 50 kHz up to the upper limit of fundamental crystals, around 20 MHz. The unit can also serve as a pulse generator by mounting a suitable capacitor in an old crystal holder.

CALIBRATED RESISTORS AND CAPACITORS

In many measuring setups, there is a need for both fixed and variable values of resistance and capacitance, where the values are known. Start with the general-purpose bridge described in Chapter 4.

It is well worthwhile to make up several switch-selected resistance and capacity banks and to measure and record the actual values of each switch setting. If much audio work is done, it is also worthwhile to make up a switch-selected inductance bank using toroids. If standard values of the 20% series are used, four decades of resistance and capacitance can be placed in a standard minibox. Values of 11, 22, 33, 44, 66, 88, 99, 110 and 132 mH can be obtained with two toroids.

Fig. 13-17. A typical auxiliary unit for laboratory work. This is a 10-turn linear precision potentiometer mounted in an insulated case.

Variable resistances are also very useful. A convenient design is shown in Fig. 13-17. This uses a miniature 10-turn potentiometer, with a multiple-turn dial; both of these units were obtained from surplus sources.

The resistance ranges to be selected will depend somewhat on the type of work being conducted. For example, if work is all on transistors, only low values of resistance will normally be needed. With FETs or vacuum tubes, high resistances are required. Practically any resistor, wirewound or carbon, is useful at audio frequencies. For radio frequencies, only carbon resistors are suitable, and these only in the low resistance values. *Allen Bradley* potentiometers, with a maximum resistance of 100 to 500Ω, will give reasonably good results at RF.

For many experiments it is also desirable to have a variable capacitance available. A possible way of securing this is to use the 3- or 4-gang tuning capacitor from an old broadcast receiver, with the sections connected in parallel. This will give capacitance variation from approximately 30 to 1000 or 1500 pF. For higher values, switched capacitors can be used.

PROBES, CLIP LEADS, AND JUMPERS

Much time can be saved by having available a good assortment of test leads, adapters, probes, jumpers, etc. A good basic assortment of leads and jumpers for general lab use would include:

20 — 15 in. jumpers with small alligator clips
5 — 36 in. jumpers with medium alligator clips
5 — 36 in. jumpers with large alligator clips
4 — 15 in. shielded leads, phono plug to alligator clip
4 — 15 in. shielded leads, phono plug to phono plug
4 — 15 in. shielded leads, phono jade to phono jack
4 — shielded leads, phono plugs to tip jacks

The lab should also have available, as a minimum:

1 — low capacity probe
1 — divider probe, 10:1, compensated for AC and DC use.
1 — RF probe

The probes may be the ones used for the lab oscilloscope. They are also useful with the VTVM and with some of the instruments described above. If the lab oscilloscope is one of the inexpensive kits with binding-post terminals, a worthwhile addition is a BNC jack, with BNC plugs on the probes. Use RG-58C/U coaxial lead for the probes.

A number of adapters are also needed. Typical ones are

- Phono jack to phono plug
- Phono plug to UHF plug
- UHF plug to BNC plug
- Banana plug to phono jack

Keep a list of the types needed as work progresses and lay in a stock on the next visit to the radio store.

Chapter 14
Be Your Own Designer

This chapter is addressed to the experimenter who has not been designing his own instruments—who has been content to use the instruments available, or perhaps has done no more than assemble one of the many commercially available kit designs.

If you have built several of the instruments described here, you should be ready to start a few such projects on your own. It really isn't difficult—it's a three-step process, involving recognition of need, establishment of an approach, and development of this approach into a workable design.

Many people seem to hang up at the first step. The usual starting, and stopping, place is, "Gee, I wish I had one of those...." Fine, but now go ahead, with the intention of finishing with one in hand.

Most designers, professional as well as amateur, have two little tricks which make design work much easier. One is a file of standard circuits—ones which have been used in the past, whose capabilities and limitations are known. The second is similar, and usually takes the form of a notebook or clipping file, stuffed with ideas. Probably most are from published designs or from the various idea columns, but others are personal notes, the result of thinking about some problem.

Having these standard circuits available encourages another trick of designers, that of thinking in terms of *black boxes*. This method ignores the details of circuits, and concentrates on what it is supposed to do, on its inputs and outputs. The circuit is represented by the familiar box of the block diagram—the black box. Details can come later.

This black-box point seems to be the good place to decide to either go ahead with an idea or drop it. If the block diagram seems to indicate that the desired performance is not too difficult to attain, that the effort is reasonable, the next step can start. The first part is expansion of the circuit to show components. Initially, don't worry about component values. Get the circuit laid out, then return and fill in component values. The file of pet circuits is a big help here, as are the practical construction handbooks, and the circuit dictionaries and handbooks. If the circuit is new, or a major change is needed, a breadboard trial may be in order. Several approach ideas may be tried this way, on paper or as hardware.

It's a good idea to hoard several circuit diagrams for each contemplated project—for a very good reason: Many amateur builders will see a published circuit and then attempt to build the item from the available information. As often as not, the circuit won't work the way the author's purportedly did. Here, most builders get discouraged and scrap the project under the assumption they made wiring errors or goofed in some interpretation of the author's suggestions. More likely, though, the artist who drew the diagram for the publications made a mistake that somehow slipped by undetected. Perhaps a couple of component values got transposed or a decimal point was incorrectly positioned. If you have several schematics available, you can compare values and spot any major discrepancies. If two similar circuits show emitter resistors of 0.47Ω or 0.33Ω, and the third shows 47Ω, you have good reason to suspect a schematic error in that third diagram. (It's even conceivable that *this* book will have schematic errors by the time the book comes off the presses.)

Most of the time a complete breadboard of the developing concept will save time and insure the best results. This can omit such functions as power supply, and often can be simplified by omission of such features as range switching, or very standard circuits such as those for metering. However, sometimes it is a good idea to do the breadboard work in what is almost a prototype, at least as far as the major elements are concerned. This is especially true for RF and high-impedance audio circuits which are sensitive to strays.

It is likely that some concept of the finished appearance of the instrument will develop during successive steps. In fact, the design seems to progress better if the end result is always kept in mind, and if some preliminary sketches of final appearance are made during black-box sketching. These sketches are a great help in picking out components for the final design.

As a help toward encouragement of beginners in undertaking circuit designs on their own, the descriptions in

Table 14-1. Adapting Circuits For New Applications.

New Instrument	Circuit Source
LF/MF/HR signal generator	o-meter
Precision sweep generator	Function generator
Analog computer	Step-gain amplifier
	Function generator
Precision pulse generator	Tone-burst generator
Shutter tester	Digital frequency meter
Dual-sweep generator	Tone-burst generator
Inductance bridge	RC(L) bridge

the previous chapters have included more comments as to limitations than are normally found in "let's build this" articles. Elimination of the deficiencies in a particular instrument makes a good design project. For example, the function generator of Chapter 9 has the deficiencies of fixed output level and high output impedance. Addition of low-impedance amplifiers, with calibrated level controls, would make a relatively simple design project, and it would give an instrument of much greater flexibility.

As additional help, each instrument description includes a discussion of principle of operation, and of use. Our purpose is to simplify use of the instrument concept in other designs, through inclusion as standard circuit elements. Some possible instruments, and a suitable circuit source, are shown in Table 14-1.

In working out designs, review of the references and other literature is most helpful. Don't forget the catalogs of the commercial instrument makers and kit suppliers. These are a great help in deciding what features to retain or omit, and such details as establishing accuracy goals, planning layouts, and even in deciding on dial calibrations and panel marking. And, of course, the catalogs are really necessary if a make-or-buy decision is to be made.

Chapter 15

Oscilloscope Improvement

Many experimenters use relatively simple oscilloscopes, perhaps to keep their investment down, perhaps because they have not gotten into the habit of using the oscilloscope as a general-purpose lab instrument. The following material describes ways to improve these old scopes, to provide some of the features of modern precision designs.

The major points covered are:

- Voltage and sweep speed calibrator
- Triggered sweep
- DC amplification
- Calibration
- Photography

The last two may be of interest to owners of precision scopes.

SAFETY FIRST

Before starting on actual work, review your safety procedures. Scopes use high voltages, and most internal layouts make no attempt to prevent contact with these. Older designs are particularly bad, in that high-capacitance filters were used in the high-voltage circuits. Take care!

GENERAL CONDITION

If your scope is one of a series, as for the common kits, it is a good idea to secure a copy of the schematic diagram for the current model. Comparison will often show simple improvements and suggest component values.

In any scope more than a few years old, it is a good bet that some maintenance is due. Recommendation: Take care of this first—it may eliminate the need for some changes, and it certainly will prevent a lot of exasperation later.

Clean the cabinet and chassis; clean and check all present tubes; make a general resistance check, to check capacitors (particularly electrolytics) for leakage, and the resistance of carbon resistors; clean and lubricate all switches and controls; check electrolytics for possible shifts in value.

If it is necessary to replace controls in the high-voltage circuits, select the replacement to have ample insulation. Many modern units are intended for low-voltage circuits only.

The timing capacitors in the sweep circuit may require change, since their leakage must be small in comparison to the 5—10 megohm timing resistor almost invariably used. Mylar, polystyrene, and silver—mica types are best for replacement.

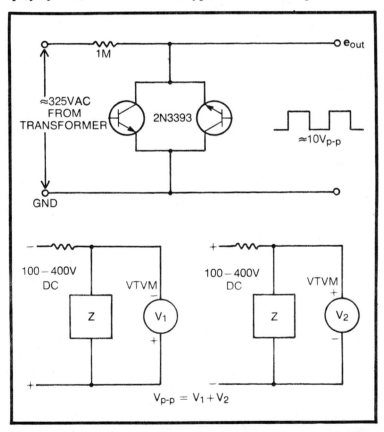

Fig. 15-1. Generation of calibrate signal: (A) circuit of square wave generator; (B) measurement of peak-to-peak amplitude using DC meter and supply.

THE CALIBRATOR

In early scopes the calibrator, if present at all, was a binding post fed from the 6.3V AC supply through a small protective resistor. Later units used a voltage divider, usually selected to give a 1V p-p signal.

It is usually easier to work with a square-wave calibration signal. If this is generated with minor precautions, it becomes a standard for time measurements. An easy way to secure this square wave is to use zener diodes fed from the high-voltage AC supply. A suitable circuit and calibration method is shown in Fig. 15-1. This gives a square wave with 8.33 msec between zero crossings, with a rise time of less than 100 μsec. The peak-to-peak amplitude is nearly independent of input and relatively insensitive to temperature. Practically any small low-leakage transistors can be used. If the wave produced is not flat-topped, use another pair.

TRIGGERED-SWEEP CIRCUITS

One of the major features of the higher priced modern scopes is the precision sweep circuit, with positive trigger control. This makes possible measurement of time and frequency to good accuracy. The better circuits allow expansion of a part of the sweep, and all allow independent choice of trigger point and sweep speed. Unwanted portions of the sweep are simply blanked.

It is not possible to provide all of these features in a simple way, but a lot can be done. The most important step is to provide a triggered sweep. This is easily done by interrupting the sweep at the end of its travel and restarting it by trigger. The principle of this is shown in Fig. 15-2 for the gas-tube sweep in older scopes. Due to the time needed to discharge the timing capacitor and to deionize the gas, the maximum frequency is limited to about 50 kHz. Sweep speeds can be fairly good, however—a microsecond or two per centimeter being possible.

Over the years the designers of low-cost scopes have made many changes in sweep circuits. One of the best developed is the *cathode-coupled multivibrator*, shown in Fig. 15-3, with provisions for sweep interruption added. Sweep frequency limit (500−2000 kHz) is much higher, and sweep speeds up to 100 nsec/cm are possible. The sweep arrangement in very old oscilloscopes might well be converted to this circuit.

DC COUPLING CIRCUITS

There does not seem to be a simple way of improving the horizontal and vertical amplifier chains appreciably. This

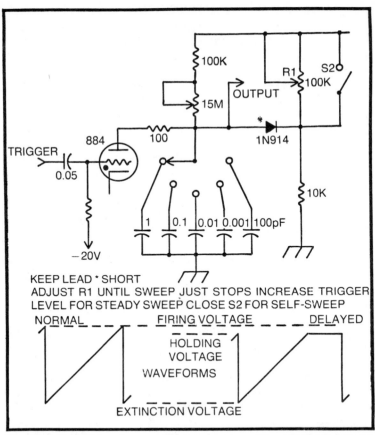

Fig. 15-2. Typical thyratron oscilloscope sweep, with added elements for trigger control.

means that the upper and lower frequency response limits established by the basic design are difficult to change. If greater range is needed, the best procedure is to use external amplifiers (as described in Chapter 6 or 7), coupling these to the deflection plates or to the grids of the output stages using AC and DC coupling as needed.

Figure 15-4 shows a typical output amplifier, and a possible direct-coupling method. In this circuit a bias is needed when DC coupling is to be used; the easiest method of supplying this is to use a series battery, as shown. In other output circuits, the centering controls are in the plate or cathode part of the circuit, and the bias battery is not needed.

Figure 15-4 also shows cross neutralization as a means of improving stability. The step may not be necessary, or only

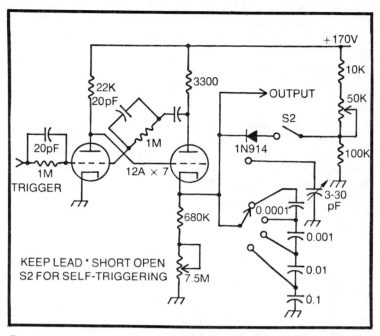

Fig. 15-3. Typical multivibrator oscilloscope sweep with added elements for trigger control.

one cross coupling capacitor may be adequate. The neutralization capacitor can be a simple *gimmick* (a length of insulated wire soldered to one plate and wrapped around the opposite grid lead).

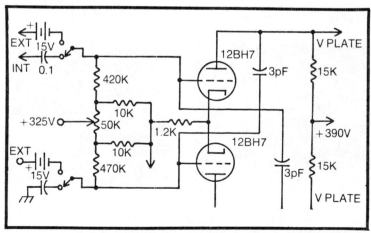

Fig. 15-4. Typical output amplifier, with added circuitry for DC coupling.

One recommended change for general-purpose scopes is to remove the amplifier peaking coils, if used. They do give an improvement of the half-power frequency, but at the expense of added attenuation at still higher frequencies. They can also cause erroneous signal presentation due to resonances and ringing.

VOLTAGE REGULATION

None of the low-cost scopes use voltage regulation. As a result, the scope is not too stable, as evidenced by change in spot position, irregular sweep amplitude, and poor calibration. Try to locate one of the voltage-regulating transformers, and operate the scope from this. An alternative is to secure partial regulation by adding a series regulator to the low-voltage circuits.

AMPLIFIER ATTENUATORS

The oldest designs of low-cost scopes used only a variable attenuator or gain control in the amplifier circuits. Later designs introduced a step attenuator in the vertical channel. If the step attenuator is not installed, it is a desirable addition. The component values of an existing or added step attenuator should be adjusted until the scale change, usually 10:1, is exact. To do this, first adjust the resistor values for 10:1 ratio for DC, using a VTVM. Then adjust compensating capacitors to give the same ratio for AC signals by feeding in a square wave or a ramp (sweep) signal, and adjusting the capacitor values for the correct pattern. The ramp signal can be obtained by connecting the vertical input to the horizontal amplifier output. In Fig. 15-5, photo A shows the effect of adjustment on

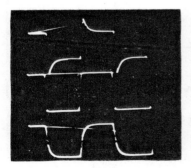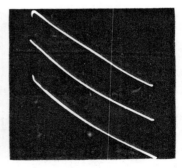

Fig. 15-5. (A) Adjustment of input attenuator capacitance compensation, using square-wave input: rounded corners—C too low; square corners—C correct; spikes—C too high. (B) Adjustment of input attenuator capacitance compensation, using scope sweep connected to vertical input: dip at end—C too low; straight—C correct; hook at end—C too high.

a ramp signal, and photo B shows the effect on a square-wave signal.

CALIBRATION

The two possibilities for calibration are *absolute* and *relative*. An absolute term would be, for example, volts per centimeter (V/cm); a relative calibration might be decibels below an arbitrary value. For the vertical amplifier and the sweep circuit, absolute calibration is best. Relative calibration is satisfactory for the horizontal amplifier.

For either method use a standardized number scale. The series 1.0, 1.2, 1.5, 2, 2.5, 3, 4, 5, 7, 10 is commonly used for this. Sometimes it is necessary to "fold" the scale, the top part reading 10, 12, 15, etc. The numbers to use can be determined as part of the calibration, as described in Chapter 3.

The controls to be calibrated, and the best calibration quantities to use for older scopes are given in Table 15-1.

The last two items in the table will need a specification of sweep-width setting, determined by the horizontal gain control. If the scope uses the older graticule marked in inches, your best bet is to change it.

SCOPE PHOTOGRAPHY

It is very convenient to have a permanent record of the scope signals found when finishing up a bit of development work. The references in the bibliography describe ways of using Polaroid cameras for this, by modifying the shutter and adding a closeup attachment plus a permanent mounting.

A recent addition to the Polaroid line is the small *Zip* camera. To adapt this type for scope use, perform the following steps:

1. Remove back.
2. Remove batteries.
3. Remove 2 bright screws from battery leads.
4. Remove ⅜ inch long black screw near top of left battery.
5. Lift off shutter assembly.

Table 15-1. Scope Controls and Their Calibration.

CONTROL	CALIBRATED IN
Vertical step attenuator	1/1, 1/10, 1/100
Vertical gain control	dB below max gain
Sweep freq. vernier	μsec/cm
Sweep freq. selector	multiplier factor

6. Remove black FH screw at lower right.
7. Lift off back plate, noting location of plastic chip.
8. Lift out flash ejector.
9. Remove black screw at top right of shutter plate.
10. Remove black screw at lower left of shutter plate.
11. Slide shutter assembly forward and up, then free.
12. Locate a point near center of small black insulator block, and about ⅛ inch away. Place tip of scratch awl here. Shutter should now stay open as long as trip is held down, and close when released. Move awl until this action is secured. Mark this point.
13. Drill for 2—56 screw, and install screw (⅜ inch long).
14. Reassemble in reverse order, omitting batteries.

To change the focal length of the lens, install a No. 3 closeup lens in front of the regular lens, which gives a lens-to-oscilloscope distance of about 10 inches and a field of view suitable for a 5-inch scope. The closeup lens may be screwed to the regular lens. Alternatively, find a plastic bottle top to fit the closeup lens, and punch a 1⅛-inch hole in it, sliding it over the regular lens.

A few sample exposures will show the proper setting of the intensity control. It seems best to use an exposure of 1 second.

Bibliography

Chapter 2

The Radio Amateurs Handbook, American Radio Relay League, Newington, Conn.

The Radio Handbook, Editors and Engineers, New Augusta, Ind.

Roubal, *Easy Way to Make PC Boards,* Popular Electronics, Oct. 1973.

Hutchinson, *Practical Photofabrication of Printed Circuit Boards,* Ham Radio, Aug. 1971.

McCoy, *Where Can I Buy the Parts?,* QST, July 1973.

VE3GFN, *For That Professional Look,* Ham Radio, March 1968.

Chapter 3

Gibson, *Test Equipment for the Radio Amateur,* RSGB, London.

Cooper, *Electronic Instrumentation and Measuring Techniques,* Prentice-Hall, Englewood, N.J. 1970.

Audio Cyclopedia, Howard W. Sams, Indianapolis 1969.

Radio Engineers Handbook, McGraw-Hill, New York 1943.

The Radio Amateurs Handbook, ARRL, Newington, Conn. 1975.

Chapter 4

Radio Engineers Handbook, McGraw-Hill, New York 1943.

Reference Data for Radio Engineers, 5th ed., Howard W. Sams, Indianapolis.

Impedance Bridges Assembled from Laboratory Parts, General Radio Co., West Concord, Mass.

Chapter 5

Radio Engineers Handbook, Mcgraw-Hill, New York 1943.

Hints and Kinks, QST, Sept. 1972.

Ideas for Design, Electronics Design, April 1971.

Chapter 6

Texas Instruments, *Technical Information on 2N2188 Applications,* Dallas.

Belcher and Victor, *General-Purpose Solid-State Preamplifier,* QST, Sept. 1968.

Chapter 7

National Linear Applications Handbook, National Semiconductor Corp. Jan. 1972.

Chapter 8

The General Radio Experimenter, Feb. 1967.
General Radio Catalog, description of 1396-B tone-burst generator.
Audio Tone-Burst Generator, in **Radio-Electronics Hobby Projects**, TAB, Blue Ridge Summit 1971.

Chapter 9

Linear Applications, National Semiconductor, Santa Clara, Jan. 1972.
Heckt, *3-Way Waveform Generator*, in **Radio Electronics Hobby Projects**, TAB, Blue Ridge Summit 1971.

Chapter 10

Cooper, *Electronic Instrumentation and Measurement Techniques*, Prentice-Hall 1970.
Reference Data for Radio Engineers, 5th ed., H.W. Sams 1968.

Chapter 11

Grillo, *A Frequency Counter with Binary-Coded Decimal Output*, QST, Aug. 1969.
Hewlett-Packard Catalog (especially editions just prior to 1972).
Secondary-Standard Frequency Comparison with TV Color Burst Frequency, Technical Correspondence, QST, Nov. 1972.

Chapter 12

The Audio Cyclopedia, Howard W. Sams, Indianapolis 1969.
General Radio and *Hewlett-Packard* catalogs, current issues.
General Radio Co., *Handbook of Noise Measurements*, West Concord, Mass.
Reference Data for Radio Engineers, 5th ed., Howard W. Sams, Indianapolis 1968.

Chapter 13

Simplified Modern Filter Design, Hayden, New York 1966.
Lancaster, *The Butterworth Filter Cookbook*, CQ, Nov.-Dec. 1966.
The Audio Cyclopedia, Howard W. Sams, Indianapolis 1969.

Chapter 15

The Radio Handbook, current ed.
Mark, *Triggered Sweep Conversion for Oscilloscopes*, QST, Dec. 1972.
Cooper, *Electronic Instrumentation and Measuring Techniques*, Prentice-Hall 1970.
Coy, *Scope Camera*, in **Radio-Electronics Hobby Projects**, TAB, Blue Ridge Summit 1971.
Dodd, *A High Quality Low-Cost Oscilloscope Camera*, QST, March 1974.

Index

238